T0133982

STATISTICAL METHODS IN PSYCHIATRY RESEARCH AND SPSS

Second Edition

STATISTICAL METHODS IN PSYCHIATRY RESEARCH AND SPSS

Second Edition

M. Venkataswamy Reddy, PhD

Apple Academic Press Inc.   Apple Academic Press Inc.
3333 Mistwell Crescent   9 Spinnaker Way
Oakville, ON L6L 0A2 Canada   Waretown, NJ 08758 USA

© 2019 by Apple Academic Press, Inc.
No claim to original U.S. Government works

International Standard Book Number-13: 978-1-77188-781-6 (Hardcover)
International Standard Book Number-13: 978-0-429-02330-9 (eBook)

Library and Archives Canada Cataloguing in Publication

Reddy, M. Venkataswamy, author
Statistical methods in psychiatry research and SPSS / M. Venkataswamy
Reddy, PhD. -- Second edition.
Includes bibliographical references and index.
Issued in print and electronic formats.
ISBN 978-1-77188-781-6 (hardcover).--ISBN 978-0-429-02330-9 (PDF)
1. Psychiatry--Research--Statistical methods. 2. SPSS (Computer file). I. Title.
RC337.R38 2019   616.89'00727   C2018-905109-4   C2018-905110-8

Library of Congress Cataloging-in-Publication Data

Names: Reddy, M. Venkataswamy, author.
Title: Statistical methods in psychiatry research and SPSS / M. Venkataswamy Reddy.
Description: Second edition. | Toronto ; New Jersey : Apple Academic Press, 2018. | Includes bibliographical references and index.
Identifiers: LCCN 2018041829 (print) | LCCN 2018042945 (ebook) | ISBN 9780429023309 (ebook) | ISBN 9781771887816 (hardcover : alk. paper)
Subjects: | MESH: SPSS (Computer file) | Psychometrics--methods | Statistics as Topic--methods | Biomedical Research--methods | Software Classification: LCC RC337 (ebook) | LCC RC337 (print) | NLM BF 39 | DDC 616.890072/7--dc23
LC record available at https://lccn.loc.gov/2018041829

ABOUT THE AUTHOR

M. Venkataswamy Reddy, PhD

M. Venkataswamy Reddy is a retired professor and former Head of the Department of Biostatistics at the National Institute of Mental Health and Neurosciences (NIMHANS), Banglore, India. He was in charge of the medical records section of NIMHANS. He has served as a member of the Board of Management of the institute. He has guided two research scholars for their PhD degrees, and they have since occupied key positions in prestigious institutes in India. He is an active life member of the Indian Psychiatric Society and the Indian Society for Medical Statistics and a member of the Indian Statistical Institute, International Biometric Society, International Epidemiological Association, and Computer Society of India. He has actively participated in several national and international conferences and has published more than 20 scientific papers. His popular textbook, titled *Statistics for Mental Health Care Research*, is widely referred to by students and academicians alike. His main contributions include the development of suitable models for mental health delivery systems in India, meta-analysis for psychiatry research and STATA, and cluster analysis for psychiatry research and SPSS. He received an MSc in statistics from the Department of Statistics at Bangalore University; a DBS (PG diploma in Biostatistics) from ICMR's Institute for Research in Medical Statistics, New Delhi; and a PhD in Biostatistics (topic: cluster formation in psychiatry with special reference to child psychiatry) from Bangalore University.

CONTENTS

LIST OF ABBREVIATIONS

AGR	annual growth rate
ANCOVA	analysis of covariance
ANOVA	analysis of variance
CDR	crude death rate
CI	confidence interval
CRD	completely randomized design
CV	coefficient of variation
DOS	duration of stay
DSD	dissociative disorder
DV	dependent variable
ECT	electroconvulsive therapy
FN	false negative
IPSS	Indian Psychiatric Survey Schedule
IQ	intelligent quotient
MANOVA	multivariate analysis of variance
NIMHANS	National Institute of Mental Health and Neurosciences
OCD	obsessive and compulsive disorder
OR	odds ratio
OTD	obsessive thought disorder
PCI	per capita income
RBD	randomized block design
RPES	rapid psychiatric examination schedule
RR	risk ratio
SE	standard error
SES	socio-economic status
SOM	somatoform
SPSS	statistical package for social sciences
TP	true positive

PREFACE TO THE FIRST EDITION

This book is best suited for professionals, teachers, and post-graduate students in the fields of psychiatry and allied mental health such as psychiatric social work, psychiatric nursing, psychiatric epidemiology, psychiatric statistics, mental health education, and clinical psychology. Keeping a watchful eye on the perfection of knowledge of these methods, this book has been prepared with a view to fill gaps in their applications and to introduce suitable data analysis software. The main objectives of writing this book include classification of fundamental statistical methods in psychiatry research, introduction to multivariate data analysis, and data analysis software presented in a precise and simple manner.

Hierarchical classification of the contents: The sequence of the 22 chapters, the sections within the chapters, the subsections within the sections, and the points listed under these subsections have been so arranged to help the professionals in classification of their knowledge in statistical methods (see Appendix II) and fill the gaps. The reader can make a glossary by referring to the meanings and definitions of the various terms given in the appendix to the main text of the book. This appendix can also provide an index.

Precise: My intension is to provide an overview of statistical methods, but the works that I have found most useful are included in the bibliography. This feature of the book helps the reader, who merely requires having his memory refreshed, in preparing a summary of statistical formula from the text of the book.

Simplicity: The fundamental statistical methods are demonstrated by means of arithmetical examples that may be reworked with pencil and paper in a matter of minutes. Most of these examples use artificial data: real data are usually more difficult to work with and can easily obscure the point being made. Care has been taken to avoid using terminology that is likely to cause difficulty to the reader.

Data analysis software: The results of the rework by the reader have to be checked by using SPSS; and in this way, the psychiatrist friendly data analysis software (SPSS) is introduced.

PREFACE TO THE SECOND EDITION

I have great pleasure to bring out the second edition of this book. In this edition, most of the chapters have undergone minor changes. Chapter 5 (Mental Health Statistics) has been expanded on, and a new Appendix III (Area, Population Size, and Density of Population of Countries) has been added with a view to help planners to provide effective mental health services and obtain accurate epidemiological information at the national and international level. Chapter 16 (Cluster Analysis) has been radically modified as these methods are most useful in psychiatry research, and Appendix IV (Classification of Child Psychiatric Disorders) has been added. These real data for illustrations are taken from my own work and publications. I hope that these models pertaining to India may be applied by policymakers and planners in other countries as well.

— **M. Venkataswamy Reddy**

ACKNOWLEDGMENT

I am grateful to Dr. P. Satishchandra, Director/Vice Chancellor, National Institute of Mental Health and Neurosciences (NIMHANS), Bangalore, India, for the support and encouragement during the preparation of this manuscript. I thank Dr. Mallikarjun B. Hanji and Dr. K. P. Suresh for rendering enormous help by going through the manuscript and assisting in the preparation of the solutions of the exercises. I thank Shri Ashish Kumar of Apple Academic Press for his full cooperation in bringing out the book in its present nice form.

This book is the first of its kind. Critical examination and suggestions to improve the book from any corner will be appreciated.

M. Venkataswamy Reddy

CHAPTER 1

PSYCHIATRIC RESEARCH

CONTENTS

1.1 RESEARCH QUESTIONS IN PSYCHIATRY

1.1.1 RECOGNITION

In recognizing cases (screening), the research questions may arise in the following areas:
1. Documentation of natural history of various psychiatric disorders/ diseases
2. Definitions of abnormal conditions

1.1.2 DIAGNOSIS

In diagnosing the recognized patients, the research questions may arise in the following areas:
1. Documentation of disease processing and mechanism
2. Disease description
3. Making diagnosis
4. Predicting the prognosis of the diagnosed patients

1.1.3 TREATMENT

In treating the diagnosed patients, the research questions may arise in the following areas:
1. Documentation of medical/therapeutic history of various psychiatric diseases
2. Planning of clinical trials; evaluation of different levels of the treatment; and providing standard measures of accuracy of various clinical procedures
3. Promotion of adequate clinical facilities
4. Determining the outcome of the treated cases

1.1.4 PREVENTION

In preventing diseases/disorders, the research questions may arise in the following areas:

1. Assessing of the state of mental health in the community and indication of the basic factors underlying this state of mental health in the community
2. Finding out the risk/causal factors associated with various psychiatric diseases
3. Planning and monitoring of mental health programs for specific population groups and evaluation of the total program of action
4. Promotion of mental health legislation

1.2 RESEARCH APPROACHES

1.2.1 SCIENTIFIC APPROACH

Scientific approach is the application of logic and objectivity in order to understand the phenomenon. When two mental health workers adapt this approach, both will arrive at the same conclusion with the same data in recognizing cases, making diagnosis of the patient, deciding on the course and method of treatment, predicting the outcome of the treatment, and in determining the cause of a mental disorder to prevent them. This is because of the standard features and steps involved in the scientific method.

1.2.1.1 FEATURES

The scientific method has several specific features such as the circularity aspect of facts and theory, humility, free inquiry, free from human bias, controlled observation, comparability, repeatability, ruling out metaphysical explanations, and the use of both inductive and deductive reasoning.

1.2.1.2 STEPS

The following five steps are involved in the scientific method: formulation of problem, formulation of hypotheses, modification of the problem/hypotheses, testing (verification or experimentation), and generalization.

1.2.1.3 SCIENCE AND PSYCHIATRY

In order to be called as a scientific discipline, the field of psychiatry has undergone the rigors of various non-scientific approaches to acquire knowledge such as the method of tenacity (blind belief), the method of authority (established belief), and the method of intuition (a priori method). Thus, psychiatry is the oldest art and newest science.

1.2.2 MILL'S CANONS

Statistics is used to determine causal factors of phenomena in experimental inquiries. John Stuart Mill presented logical devices as short rules under the title "methods of experimental inquiry." These rules indicate five canons.

1.2.2.1 METHOD OF DIFFERENCE

This consists in making use of two groups of subjects equal in all respects and do something to one of the groups (experimental group) and not doing anything to the other group (control group). If a change takes place in some dependent variable in the experimental group, but it does not take place in the control group, then the change in the dependent variable is attributed to the manipulation in the experimental group.

Example: Let us suppose that 60 schizophrenic patients are randomly divided into two groups of 30 patients in each group. One group is given educational input along with the treatment, and the other group is given the treatment without additional educational input. The educational input group has demonstrated better understanding of the illness which led to better follow-ups and management.

1.2.2.2 METHOD OF AGREEMENT

Whenever the researcher observes the occurrence of phenomenon, he must notice presence of specific independent and dependent variables. This consists in observing the occurrence of the phenomenon in each time where a specific independent variable presents, a specific dependent variable occurs. The researcher continues to observe the occurrence of the phenomenon with different combinations of his independent variable until he ascertains that there is only one specific independent variable that always presents whenever a specific dependent variable occurred. Then, he would infer that the independent variable that was common to the occurrence is related to the occurrence of the dependent variable.

Example: The case history records of 28 conduct disorder children revealed that all of them had either inadequate parental control or family over involvement. By applying the method of agreement, it is possible to state a relationship between the occurrence of conduct disorder and the presence of inadequate parental control or family over involved.

1.2.2.3   JOINT METHOD OF DIFFERENCE AND AGREEMENT

This method consists in testing to see that if in two or more instances, the phenomenon has only one factor in common and then test to see that if in two or more instances where that common factor is absent, the phenomenon does not occur. The researcher concludes that the common factor is related to the occurrence of the phenomenon.

Example: The case history records of 200 child guidance clinic children revealed that all the 23 children with hyperkinetic disorder had abnormal attention and concentration, and only two of the remaining children had this abnormality. Hence, it can be said that certain relationship may exist between the occurrence of hyperkinetic disorders and the presence of abnormal attention/concentration in children.

1.2.2.4   METHOD OF CONCOMITANT VARIATION

This method consists in recording the variation of both the independent variable and the dependent variable; and if the dependent variable varies in any manner whenever the independent variable varies in some particular manner, the experimenter concludes that the two variables are related.

Example: The findings of mental morbidity studies carried out in India revealed several relationships between the prevalence of mental/behavioral disorders and several biosocial variables. The findings further indicate that the risk of mental and behavioral disorders of a person decreases with increased socio-economic status, increases gradually up to 40 years of age and decreases thereafter, and increases with both the size of his family, urbanization of his locality, etc.

1.2.2.5   METHOD OF RESIDUES

This method consists in attempting to determine, through experimentation and deduction, the phenomena that are due to the effect of the presence of specific and identifiable independent variables. The researcher continues to ascertain such information until the relationship between one dependent variable and one independent variable is unknown in the situation. Then, the investigator infers that this remaining independent variable is related to the remaining dependent variable.

Example: A series of five drugs are administered to a group of patients and the responses are noted down. Then, each drug is withdrawn every time from least important to most important in order to find out at which withdrawal of the drug the patients stop improving or regressing.

1.2.3   QUALITATIVE RESEARCH

Qualitative research involves collecting qualitative data by way of in-depth interviews, open-ended questions, observations, and field notes. The researcher is the primary source of data collection, and the data could be collected in the form of words, images, and pattern. Data analysis involves searching for pattern, themes, and holistic features. Results of such research are context specific, and reporting takes the form of a narrative with contextual descriptions and direct quotation from researchers.

1.2.4   QUANTITATIVE RESEARCH

Quantitative research involves collecting quantitative data based on precise measurements using structured, reliable, and validated data collection instruments or through archival data sources. The nature of the data is in the form of variables. Based on the purpose of the study, the studies of quantitative research can be broadly classified as observational studies and experimental studies.

1.3 PROTOCOL WRITING FOR QUANTITATIVE STUDIES

1.3.1 THE RESEARCH QUESTION

1.3.1.1 REVIEW OF LITERATURE

A research question/problem arises in two occasions:first, to develop a theory to explain an event that occurred, and none of the existing theories explained it so far and second, to modify the existing theory so as to suit the new event. In order to identify the problem, the scientist must have good acquaintance with the area in which he has to identify the problem. Thus, he has to review the relevant literature (standard authors' work).

1.3.1.2 NEED FOR THE STUDY

The need for the study is an important consideration as it specifies the priority to be given and the budget that is required for the study.

1.3.1.3 AIMS AND OBJECTIVES

The aim of the study is to find out the truth that is hidden and that has not been discovered so far. The objectives of the study are to discover answers to questions using scientific procedures. According to the nature of the objectives, the studies may be classified as exploratory to gain familiarity with a phenomenon, descriptive to portray accurately the characteristics of a particular group, hypothesis testing to test a hypothesis of a causal relationship between variables, or diagnostic to determine the frequency with which something is associated with something else.

1.3.2 TYPE OF STUDY

The protocol writer has to specify the type of study, whether it is observational (case-series, cross-sectional, or longitudinal) or experimental (informal or formal).

1.3.3 PLAN OF STUDY

1.3.3.1 POPULATION

The term "population" has wider meaning in any research activity. It is the aggregate of units of observation about which certain information is required. When the investigation is carried out for all the units in the population, it is called census enumeration. The statistical constants based on population values are known as parameters. In psychiatric research, we are mainly concerned with populations such as the following:

1. General population of a defined community
2. Psychiatric patients at a particular point of time in the community
3. New psychiatric patients during a particular time period in the community
4. Out-patient registrations in a mental health delivery system during a time period (say 1 year)
5. Admitted patients in a mental health institution/general hospital psychiatric unit during a time period (say 1year)
6. All in-patients at a particular point of time in a mental health institute/general hospital psychiatric unit
7. Discharged patients during a particular time period.

1.3.3.2 VARIABLES

The characteristics on which the observations are made are known as variables such as age, sex, diagnosis, etc.

1.3.3.3 STANDARDIZATION

All the terms/categories used in the study have to be clearly defined and standardized. For example, an urban locality may be defined as one with closed drainage facility, semi-urban as one with open drainage facility, and rural as one with no drainage facility. In mental health information system, a death in a hospital can be conveniently defined as a specific category of discharge.

1.3.3.4   EXPERIMENTAL SETTINGS (IF ANY)

The experimental setup (if any) is where a variable may be changed so that changes in the final output may be compared with the control setup. The control setup will be the basis of all measurements and a means of comparison.

1.3.3.5   RECORD FORM

In both observational studies and experimental studies, the investigator collects data by using a *proforma* that may be a questionnaire, schedule, or record form. In questionnaire approach, the respondents fill the *proforma* and return them or send them by post. In the schedule approach, the investigator asks specific questions and fills the *proforma*. The record forms are used to extract secondary data. Before the finalization of *proforma*, it is advantageous to carry out a pilot study on a small number of units. The results of the pilot study may be used to determine the sample size required for major studies.

1.3.3.6   RESPONSE ERROR

Response errors are the wrong or biased answers. The quality of the data collected may be assessed by cross checking. In experimental studies, the response errors are referred to as experimental errors.

1.3.3.7   NON-RESPONSE ERROR

Non-response errors mean failure to measure some of the units in the selected sample. Such errors may be minimized through adequate planning, training, monitoring, and supervision.

1.3.4 PLAN OF ANALYSIS

1.3.4.1 STATISTICAL METHODS

Statistical methods are the tools of science to deal with mass data. These are the mathematical devices of use in measuring variables/traits; discovering certain differences, relationships, probable predictions/trends, occurrence of events by time, and decomposition of seasonal variation; discovering distinct groups; identification of individuals; reduction of data, and synthesizing the results of similar studies. Based on probability scale, they help the researchers to decide whether these differences, relationships, trends, etc., are large enough to be considered as significant (real) or due to chance factor.

1.3.4.2 DATABASE AND DATA ANALYSIS SOFTWARE

A store of information often held on a computer as a database is stored as a number of records or files, each of which usually contains entries under a number of headings or fields. For example, a psychiatrist keeps a database under the following headings: name of the patient, age, sex, date of visit, symptoms, diagnosis, treatment, etc. The ability of a computer to cross-reference within a database makes it a particularly efficient form for storing and handling data. It could, for example, be made to search the database for all patients suffering from schizophrenia.

The ready-made set of instructions called package have been developed, which can be used to carry out statistical analysis of the data. The statistical package for social sciences SPSS (Chapters 21 and 22) are the commonly used data analysis software in psychiatric research and allied fields.

1.3.5 REPORTING THE RESULTS

Before reporting the results, the study has to be evaluated and checked to see whether the results are correctly interpreted. The reporting of the results of scientific studies is different from those of the technical reports. The report must be reasonably brief, with tables and graphs when necessary, and reading the report should be a stimulating and satisfying experience.

1.4 VARIABLES IN PSYCHIATRY

1.4.1 *NATURE OF SOME OF THE VARIABLES IN PSYCHIATRY*

(a) Demographic variables: Age, Sex, Marital status, etc.
(b) Socio-cultural variables: Religion, Caste, Region, Domicile, etc.
(c) Socio-economic variables: Education, Occupation, Income, etc.
(d) Family variables: Family type, Family structure, Family size, etc.
(e) Symptoms: Attention difficulty, hallucinations, Amnesia, Delusions, etc.
(f) Causal factors: Genetics, Environmental, Physical, etc.
(g) Faculties of the mind: Cognition, Emotion, Volition, etc.
(h) Neurosis: Hysterical, Anxiety, Obsessive compulsive, Phobia, Depression, etc
(i) Personality disorders: Personality pattern disorders, Sexual perversions, etc.
(j) Child psychiatric disturbances: Behavioral disorders, Habit disorders, Psychiatric disorders, etc.
(k) Cultural-bound Syndromes: Acute anxiety syndromes, Dissociation syndromes, etc.
(l) Treatment: Psychotherapy, Psychotropic drugs, Electro-convulsive therapy, etc.
(m) Hospital: Hospitalization, Duration of stay, etc.
(n) Outcome: Type of discharge, Result of treatment, etc.

1.4.2 *QUALITATIVE AND QUANTITATIVE VARIABLES*

A qualitative variable (sex, religion, diagnosis, etc.) is one in which the variables differ in kind rather than in magnitude. A qualitative variable may be dichotomous such as sex (male and female) or polychotomous such as religion (Hindu, Muslim, Christian, etc.). The categories (domain) of some of the important variables are sex (male, female, etc.), marital status (unmarried, married, widow/er, divorced, separated, remarried, etc.), religion (Hindu, Muslim, Christian, etc.), domicile (rural, semi-urban, urban), family type (nuclear, joint, extended, living together), and diagnosis (ICD-11).A quantitative variable (age, income, etc.) is one in which the variates differ in magnitude.

1.4.3   LEVELS OF MEASUREMENT OF VARIABLES

Qualitative (categorical) variables can be classified as nominal and ordinal variables according to the level of measurement. The quantitative variables (discrete or continuous) can be classified as interval-scale variables and ratio-scale variables.

1.4.3.1   NOMINAL LEVEL

Measurement in which a name is assigned to each observation belongs to the nominal scale of measurement. The variables such as sex, religion, diagnosis, etc., are measured in nominal scale of measurement.

1.4.3.2   ORDINAL LEVEL

The ordinal scale differs from the nominal scale in that it ranks the different categories specified in the scale in terms of a graded order. Variables such as severity of mental retardation (borderline, mild, moderate, severe, profound) and socio-economic status (low, middle, high) are measured in ordinal level/scale. Ranked variable is a special kind of ordinal variable in which the individual observations can be put in order from smallest to largest, even though the exact values are known.

1.4.3.3   INTERVAL LEVEL

The interval scale has all the characteristics of an ordinal scale. In addition, the distances between the successive scale points are of equal size. Variables such as intelligent quotient (IQ), achievement test scores, etc., are measured in interval scale of measurement. Here, zero point and the unit of measurement are arbitrary.

1.4.3.4   RATIO LEVEL

A ratio scale has all the characteristics of an interval scale; and in addition, it has an absolute zero point. Variables such as height, weight, etc., are measured in ratio scale. The scores obtained on psychosocial variables may be treated as if they are measured in ratio scale on the assumption that no serious errors will be incurred.

1.4.4   FURTHER TYPES OF VARIABLES

1.4.4.1   INDEPENDENT AND DEPENDENT VARIABLES

In psychiatric research, an independent variable is the factor that is chosen or manipulated by the experimenter in order to discover the effect on the dependent variable. A dependent variable is the factor that is observed or measured by the investigator—the outcome of the study.

1.4.4.2   COVARIATES

Covariates are interval-level independent variable. A covariate is a variable that is possibly predictive of the outcome under study. A covariate may be of direct interact or it may be a confounding one.

1.4.4.3   CONCOMITANT VARIABLES

A variable that is observed in a statistical experiment but that is not specially measured is called a concomitant variable. It is sometimes necessary to correct for concomitant variable in order to prevent distortion of the results of experiment or survey.

1.4.4.4   FACTORS/DIMENSIONS

Factors are nominal-level or ordinal-level independent variables.

1.4.5 DISCRETE AND CONTINUOUS DATA

Discrete data can only have specified values. The discrete data may be nominal data (male, female data), ordinal data (mild, moderate, severe data), ranking data, or whole number data. Continuous data are often referred to as measurement data. Continuous data can have any numerical value such as the height of a person.

1.5 STATISTICAL METHODS IN PSYCHIATRIC RESEARCH

The statistical methods selected for a particular study depend on the objectives of the study and the type of data collected for the purpose. These statistical methods can be broadly classified according to the number of variables that are simultaneously available for analysis (univariate statistical methods, bivariate statistical methods, and multivariate statistical methods).

1.5.1 ORGANIZATION AND COLLECTION OF DATA

1.5.1.1 OBSERVATIONAL STUDIES (CHAPTER 2)

Observational studies collect data from an existing situation.

1.5.1.2 EXPERIMENTAL STUDIES (CHAPTER 3)

Experimental studies are ones in which the investigator deliberately sets one or more factors at a specified level.

1.5.2 DESCRIPTIVE STATISTICS

1.5.2.1 ONE-VARIABLE DESCRIPTIVE STATISTICS (CHAPTER 4)

These methods portray accurately the characteristics such as the type of frequency distribution of observed values and summarization figures of the characteristics of the univariate data. Thus, it helps in gaining familiarity with the situations and phenomena.

1.5.2.2   MENTAL HEALTH STATISTICS (CHAPTER 5)

These are the descriptive statistics in the field of psychiatric service and research.

1.5.3   BASIS OF STATISTICAL INFERENCE

1.5.3.1   PROBABILITY AND PROBABILITY DISTRIBUTIONS (CHAPTER 6)

The probability scale is the basis of statistical inference. It includes both the estimation of parameters and test of significance of hypotheses. The descriptive statistics forms the material for statistical inference.

1.5.3.2   SAMPLING THEORY AND METHODS (CHAPTER 7)

These are the important steps in any research project as it is rarely practical, efficient, or ethical to study the whole population. The aim of random sampling approaches is to draw a representative sample from the population, so that the results of studying the sample can then be generalized back to the population. The selection of an appropriate method depends on the aim and objectives of the study and on the nature of the population to be studied.

1.5.3.3   BASIC ELEMENTS OF STATISTICAL INFERENCE (CHAPTER 8)

These elements include the estimation of parameters (criteria for selecting a suitable point estimators and interval estimators, and determining the sample size for drawing valid conclusions, etc.) and test of significance of hypotheses (formulation of hypotheses, level of significance, test statistics, etc.).

1.5.4   TESTS OF SIGNIFICANCE OF HYPOTHESES

1.5.4.1   PARAMETRIC TESTS OF SIGNIFICANCE (CHAPTER 9)

The parametric tests of significance are the test of hypothesis which assumes that, the population is particular type of distribution (normal distribution) and test the value of one of the parameters. The parameter tested depends on the distribution. Examples: The normal test in the t-test for population mean, the use of correlation coefficient for the population correlation and the chi-square test for the population variance. These include the tests of significance of difference between means (arithmetic means) of groups on a variable facilitate whether the difference is significant (real) or due to merely a chance factor.

1.5.4.2   EXPERIMENTAL DATA ANALYSIS: ANOVA (CHAPTER 10)

These include the analysis of variance (ANOVA) tests of significance. These are the extension of parametric tests of significance for more than two groups.

1.5.4.3   NON-PARAMETRIC TESTS OF SIGNIFICANCE (CHAPTER 11)

These are the methods of inference in which there are no assumptions made about the nature, shape, or form of the population from which the data are obtained. They are preferred when the sample size is small and for qualitative variable data including ranking data or order data.

1.5.5   CORRELATIONAL VARIABLES DATA ANALYSIS

1.5.5.1   CORRELATIONAL ANALYSIS AND REGRESSION ANALYSIS (CHAPTER 12)

The correlation analysis measures the strength of correlation/association between variables and then test whether these correlations are significant or due to chance factor. On the contrary, the regression analysis fits functional relationship between dependent variable (e.g., the outcome of the treatment) and the independent variables (e.g., the type of treatment, etc.), and therefore it facilitates in predicting the value of the dependent variable based on the values of the independent variables.

1.5.5.2   RELIABILITY ANALYSIS AND VALIDITY ANALYSIS (CHAPTER 13)

Reliability is the consistency of a set of measurements or of a measuring instrument, which is often used to describe a test. Instrument validity is the degree to which an instrument measures what it is intended to measure.

1.5.5.3   SURVIVAL ANALYSIS AND TIME SERIES ANALYSIS (CHAPTER 14)

There are two aspects in time-related data analysis. Survival analysis deals with the estimation of probability (chance) of occurrence of events within a specified time period or estimation of time period for an event to occur. The time series analysis determines the influence of time components such as the trend, seasonal variation, and cyclical component on the occurrence of events, and thereby forecast the events.

1.5.6   MULTIVARIATE DATA ANALYSIS

1.5.6.1   MULTIVARIATE STATISTICAL METHODS (CHAPTER 15)

These methods simultaneously analyze more than two variables. These methods take into account the various correlations of the variables, and hence they give more realistic results.

1.5.6.2   CLUSTER ANALYSIS (CHAPTER 16)

These methods form groups of similar individuals (patients) and then establish the characteristics of the groups for the purpose of a classificatory system (ICD, DSM) or for any other research goals.

1.5.6.3   DISCRIMINANT ANALYSIS (CHAPTER 17)

These methods determine a set of characteristics that can significantly differentiate between the groups and in the process allocate individual
objects (patients) into established classes (diagnostic categories) on the
basis of specific criteria.

1.5.6.4   FACTOR ANALYSIS (CHAPTER 18)

Factor analysis involves grouping of similar variables (test items, signs,
symptoms, etc.) to few factors or dimensions. The socio-economic status
is based on education, occupation, and income. Thus, the factor analysis
involves data reduction. The correlations between variables form the basic
material for this purpose.

1.5.7   MULTI-STUDY DATA-POINTS ANALYSIS

1.5.7.1   META-ANALYSIS (CHAPTER 19)

The meta-analysis synthesizes (combine systematically) the results of
similar but independent studies on a subject whenever several studies on a
subject have conflicting conclusions.

1.6 STEPS IN WRITING A PROTOCOL FOR QUANTITATIVE STUDIES

1. Introduction	(a)	Formulation of Problem
	(b)	Review of Literature
	(c)	Need for the Study
2. Aims and Objectives	(a)	Primary
	(b)	Secondary
3. Plan of Study	(a)	Population to be Studied: Criteria for Inclusion/Exclusion
	(b)	Investigations to be Made
	(c)	Standardization of Terms
	(d)	Census/Sampling
	(e)	Experimental Settings
	(f)	Proforma – Pilot Study
	(g)	Personnel – Their Duties
	(h)	Quality of Data Collected: Cross Checking, etc.
4. Plan of Analysis	(a)	Manual/Electronic Data Processing
	(b)	Statistical Methods for:
		(i) One-Variable Descriptive Statistics
		(ii) Statistical Inference
		(iii) Bivariate Data Analysis
		(iv) Multivariate Data Analysis
		(v) Meta-Analysis
	(c)	Report Writing
5. Budget		

KEYWORDS

- **Analysis of variance**
- **Dependent variable**
- **Hypothesis testing**
- **Independent variable**
- **Intelligent quotient**
- **Psychiatric disorders**
- **SPSS**

CHAPTER 2

OBSERVATIONAL STUDIES

CONTENTS

2.1   CASE-SERIES ANALYSIS

A case-series analysis report is a simple descriptive account of investigating characteristics observed in a group of registered patients.

2.1.1   OBSERVATIONAL UNITS

Observational units in a case-series analysis are the registered psychiatric patients in a mental health delivery system.

2.1.2   VARIABLES

The variables that are to be included in case-series analysis depend on the purpose of the study and the particular mental health delivery system. However, it is necessary to include important basic variables such as age, sex, and diagnosis.

2.1.3   A PROFORMA TO OBTAIN BASIC DATA OF PSYCHIATRIC PATIENTS

1. Name of the Patient:			
2. Date of Registration:			| | | | | | |
3. Registration Number of the Patient:			| | | | | | |
4. Age (completed years):			| |
5. Gender	: 1. Male	2. Female	|
6. Marital Status	: 1. Unmarried	2. Divorced	|
	3. Married	4. Separated	
	5. Widower	6. Remarried	
7. Religion	: 1. Hindu	2. Christian	|
	3. Muslim	4. Others	
8. Place of Residence	: 01 - Districts of the State		|

	50 - Neighboring States		
	60 - All other States		
	61 - All other countries		
9. Domicile	1. Urban	2. Rural	
	3. Semi-Urban		
10. Educational	: 1. Illiterate	6. Diploma / Certificate	
	2. Literate / Primary	7. Graduate	
	3. Middle	8. Post Graduate / Professional	
	4. Secondary	9. Age : Below 7 years	
	5. Higher Secondary / PUC		

11. Occupation of the Patient (codes given separately) :

12. Income per Month   :

13. Duration of Illness   :

First two columns for years

Middle two columns for month

Last two columns for days

14. Duration of Stay in Days (in-patient only):

15. Follow-up Attendance (for out-patients only):

16. Main Diagnosis (ICD – 11):

17. Subsidiary Diagnosis (if any):

18. Results of Treatment (% improvement):

2.1.4 RECORD FORM

In the record form, the rows represent the patients and the columns indicate the variables. Appendix I illustrates a typical record form to collect data of registered psychiatric patients at NIMHANS hospital.

The case-series analysis is vital because of its descriptive role as precursors to other studies. The observations from such studies may be extremely useful to investigator in designing a study to evaluate cause or explanations of the observations. They should be viewed as hypotheses generating and not as causative. The appropriate statistical methods to analyze the data in case-series records include one variable descriptive statistics.

2.2 CROSS-SECTIONAL STUDIES

A cross-sectional study collects data on a group of subjects at some fixed time, such as all in-patients (census) in a mental health institute and prevalence morbidity survey in a defined community.

2.2.1 HOSPITAL IN-PATIENTS CENSUS

2.2.1.1 OBSERVATIONAL UNITS

The observational units for in-patients census are all the in-patients in a mental hospital/psychiatric unit on a particular day (say July 1 of the year).

2.2.1.2 VARIABLES

The variables to be included in such studies depend on the type of hospital/ unit. However, it is necessary to include important variables such as age, sex, diagnosis, mode of admission status (voluntary, certified, observational, criminal record), and duration of stay in the hospital.

2.2.1.3 RECORD FORM

The record form includes the identification of particulars such as ward, admission number, and initials of the patient.

It is necessary to conduct in-patient census in mental health institute (hospital). The data analysis consists of descriptive statistics according to duration of stay.

2.2.2 MENTAL MORBIDITY STUDIES

2.2.2.1 OBSERVATIONAL UNITS

In prevalence morbidity studies, the sampling unit is the family, and the observational units are all the persons including psychiatric cases in the selected families.

2.2.2.2 VARIABLES

The variables to be studied are specified in the record form/schedules used for the survey. Four schedules are prepared to collect the data.

2.2.2.3 HOUSEHOLD INFORMATION SCHEDULE

This records the data connected with family structure, family size, age of each member of the family, their marital status, education, housing, and identification of the family.

2.2.2.4 SOCIO-ECONOMIC STATUS SCHEDULE

This is used to determine socio-economic status (SES) of each family. The schedule developed by Pareek and Trivedi is standardized for use in rural areas of India. The schedule developed by Kuppuswamy is used for urban sectors.

2.2.2.5 CASE DETECTION SCHEDULE

This contains questions that lead to the identification of all possible mental illness. The sensitivity and specificity of the screening scale to be used in the field settings has to be assessed before using it. The Indian Psychiatric-Survey Schedule (IPSS) prepared by Kapur et al., and the rapid psychiatric examination schedule (RPES) prepared by Sharma are the ones that are most commonly used.

2.2.2.6   CASE RECORD SCHEDULE

This gives all relevant information about the case detected through the case detection schedule and records the findings of examination and final diagnosis. For an operational definition of a case, ICD-11 can be taken as a model. The statistical methods in mental morbidity studies include the prevalence rates for all the major diagnostics categories according to important variables included in the study.

2.3   LONGITUDINAL STUDIES

2.3.1   RETROSPECTIVE STUDIES

Schematic representation of retrospective study design is given in Figure 2.1.

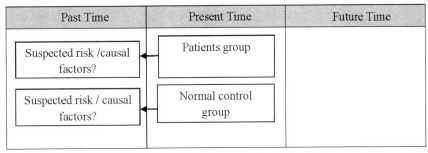

FIGURE 2.1   Schematic representation of retrospective study design.

The cases in retrospective studies (case–control studies) are individuals selected on the basis of some disease/disorder, and the controls are subjects without the disease/disorder. The history of previous events of both cases and controls are analyzed in an attempt to identify a characteristic or risk factor present in the cases histories but not in the controls histories.

2.3.1.1   PATIENT GROUP AND NORMAL CONTROL GROUP

The investigator sometimes uses matching to associate controls with cases on characteristics such as age, sex, social, and economic status. If an investigator feels that some characteristics are so important that an imbalance between the two groups of patients would affect any conclusion, then he should employ matching. This process ensures that groups will be similar with respect to important characteristics that may otherwise cloud or confound the conclusion. Sometimes, it is recommended to use of the following two controls: one control group similar in some way to the case (e.g., having been hospitalized during the same period of time) and the other control group of healthy subjects.

2.3.1.2 SUSPECTED RISK/CAUSAL FACTORS

These are the factors associated with the psychiatric disorders including genetic, environmental, and so on.

2.3.1.3 RECORD FORM

A typical record form is as given below:

Patient /Subject Number	Disease/ (Present/Absent)	Age	Sex	Ses	-	Suspected Risk/Causal Factor (Present/Absent)
1						
2						
-						

The results of a retrospective study may point to the existence of a real association between a disease and some etiological factor, but they can never prove cause-and-effect relationship. The case–control studies are especially appropriate for studying disease or events, which develop over a long time, and for investigating a preliminary hypothesis. They are generally the quickest and least-expensive studies to understand and are ideal for investigators who need to obtain some preliminary data before writing a proposal for a more complete, expensive, and time-consuming studies. In case–control studies, an estimate of relative risk known as odds ratio (OR) is calculated.

2.3.2 PROSPECTIVE STUDIES

The prospective studies (cohort studies, incidence studies, follow-up studies) select two groups based on exposure. A cohort is a group of people who have something in common and who remain part of a group over an extended time. In medicine, the subjects in cohort studies are selected by some defining characteristic to suspect of being a precursor to or risk factor for a disease. Schematic representation of prospective study design is given in Figure 2.2.

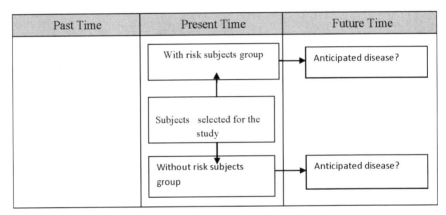

FIGURE 2.2   Schematic representation of prospective study design.

2.3.2.1 WITH RISK SUBJECTS AND WITHOUT RISK SUBJECTS

Researchers select subjects at the onset of the studies and then determine whether they have the risk factor or have been exposed. All subjects are followed over a certain period to observe the effect of the risk factor or exposure (the anticipated disease).

2.3.2.2 ANTICIPATED DISEASE

The anticipated disease/disorder may be any of the mental and behavioral disorder.

2.3.2.3   RECORD FORM

A typical record form is given below:

Subject Number	Risk Factor (Present/Absent)	Age	Sex	-	Anticipated Disease (Present/Absent)
1	-	-	-	-	-
2	-	-	-	-	-
-	-	-	-	-	-

The cohort design permits absolute and relative comparisons of disease incidence among the exposed and the unexposed. These studies have great value when a precise hypothesis has already been formulated, and it is desirable to obtain clear-cut evidence to support or refute it. When the disease is rare, a prospective study may not be possible because the required minimum number of cases will not be available from the population. They also make it difficult for investigators to argue causation because other events occurring in the intervening period may have affected the outcome. Cohort studies that require a long time to complete are especially vulnerable to problems associated with patient follow-ups, particular patient attrition (patients stop participating in the study), and patients' migration. When the data are in the form of counts, the strength of the correlation between the risk factor and the disease may be expressed as risk ratio (RR).

KEYWORDS

- **Case-series analysis**
- **Case–control studies**
- **In-patients census**
- **NIMHANS**

CHAPTER 3

EXPERIMENTAL STUDIES

CONTENTS

3.1 BASIC ELEMENTS OF EXPERIMENTS

3.1.1 EXPERIMENTAL ERROR

There are experimental errors that include all types of extraneous variation due to intrinsic variability of experimental material, lack of uniformity in the methodology of conducting the experiment, and lack of representation of the sample to the population.

3.1.2 BASIC PRINCIPLES OF EXPERIMENTS

3.1.2.1 REPLICATION

An experimenter resorts to the principle of replication in order to average out the influence of the chance factors on different experimental units.

3.1.2.2 RANDOMIZATION

The experimenter resorts to the principle of randomization to provide a logical basis for validity of the statistical tests of significance.

3.1.2.3 LOCAL CONTROL

The experimenter may resort to the principle of local control to reduce the experimental errors by dividing the heterogeneous experimental material into blocks consisting of homogeneous units.

3.1.3 LABORATORY EXPERIMENTS AND CLINICAL EXPERIMENTS

A laboratory experiment takes place in laboratory where experimental manipulation is facilitated. A clinical study takes place in the settings of clinical medicine. Experimental studies in medicine that involve humans are called clinical trials because their purpose is to draw conclusions about a particular procedure or treatment. Thus, clinical trial is a carefully and ethically designed experiment with the aim of answering some precisely framed question.

3.1.4   THERAPEUTIC TRIALS AND PROPHYLACTIC TRIALS

Therapeutic trials are the test of remedies on a disease or a disorder that is followed whenever some improved technique is reported. Such tests of treatment methods differ little from the regular clinical procedures. The differences seem largely a matter of degree of attention to details of whom to test, how to treat, and how to measure the effect of treatment.

The prophylactic trials attempt to present some diseases from occurring by applying a treatment to susceptible persons before that disease has appeared. These trials in medical practice are more convenient to run when the case load of patients at risk is high, the risk factor is not low, and the time between cause and effect is short. The basic principles of these trials are the same as that of the therapeutic trials. In prophylactic trials, the susceptible people are divided into two groups: one group is given the prophylactic and the other is not (control). These groups are then observed over the same period to see if the protected group suffered a lower incidence of the particular disease than the unprotected. These trials have been carried out particularly in infectious diseases.

3.1.5   UNCONTROLLED AND CONTROLLED TRIALS

Uncontrolled trials are studies in which the investigators experiment with the experimental drug (therapy, procedure), but the treatment is not compared with another treatment, at least not formally. The points to be considered are the treatment, patients group, measurement of outcome, record form, and the data analysis.

Controlled trials are studies in which the experimental drug or procedure is compared with another drug or procedure: sometimes a placebo and sometimes the previously accepted treatment. The purpose of an experiment is to determine whether the intervention (treatment) makes a difference. Studies with controls are much more likely than those without controls to detect whether the difference is due to the experimental treatment or to some other factor. Thus, controlled studies are viewed as having far greater validity in medicine than uncontrolled studies. The common

way a trial can be controlled is to have two groups of subjects: one that receives the experimental procedure (experimental group) and the other that
receives the placebo or standard procedure (control group). The experimental and control groups should be treated alike in all the ways except
for the procedures itself so that any differences between the groups will be
done to the procedure and not to other factors.

3.2 PARALLEL CONTROL CLINICAL TRIALS

Schematic representation of parallel control clinical trial design is given
in Figure 3.1.

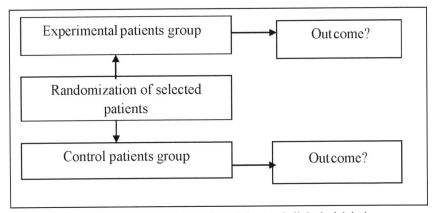

FIGURE 3.1   Schematic representation of parallel control clinical trial design.

3.2.1 TREATMENT AND DISEASE

When a treatment is already in existence, the new form of treatment should
normally be compared with the established one. As a first step, the eligible
patients from the available patients are selected as per the medical and
ethical considerations. After the procedure, a list of volunteers must be
short listed. In order to overcome the ethical problem, the volunteers must
be informed that they may be assigned to either the experimental or the
control group and a written consent of the subjects to participate must be
obtained.

3.2.2   EXPERIMENTAL PATIENTS GROUP AND CONTROL PATIENTS GROUP

3.2.2.1   RANDOMIZATION

The volunteers are stratified according to the objectives of the trials. In each group, allot the treatment and controls at random in such a way that the groups are initially equivalent in all respects relevant to the inquiry. The randomized controlled trial is the epitome of all research designs because it provides the strongest evidence for concluding causation. It provides the best insurance that the results were due to the intervention.

3.2.2.2   BLINDING

To reduce the bias, the experimenter may prefer blind trials. In a single-blind study, only the investigator knows of what is administered to each subject. In a double-blind study, neither the subjects nor the investigators know the identity of what is administered.

3.2.3   MODE OF ADMINISTRATION OF THERAPY

In the case of new treatment, the information is at first scanty. The testing may reveal some of the dangers of side effects of the drug. We may choose one dose of a drug out of many and vary the interval of its administration, given it by different routines for different time periods, and so on. Identical procedures should be followed for the control group also.

3.2.4   RECORD FORM

In observing the findings, standard record forms must be drawn up and uniformity in completing must be maintained in both the groups. Every departure from the design of the experiment lowers its efficiency to some extent. The deviations from the procedure laid down may be due to the removal of the patients due to serious side effects, deaths, deterioration of conditions, and so on. These deviations must be considered in the standard analysis of data and valid inferences have to be made.

3.3 FURTHER CONTROL CLINICAL TRIALS

3.3.1 SELF-CONTROL TRIALS

The self-control experimental design (cross-over design) uses two groups of patients: one group is assigned to the experimental treatment and the other group is assigned to the placebo or control treatment. After a time, the experimental treatment and placebo are withdrawn from both groups for a washout period. During the washout period, the patients general-ly do not receive any treatment. The groups are then given the alternate treatment; that is, the first group now receives the placebo and the second group receives the experimental treatment. Schematic representation of self-control trial is given in Figure 3.2.

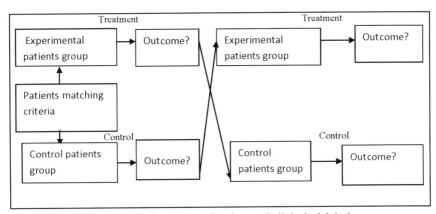

FIGURE 3.2   Schematic representation of self-control clinical trial design.

3.3.2 HISTORICAL CONTROL TRIALS

One can also undertake a cohort study by using information collected in the past and kept in records or files.

Example: We may assess study outcome of 1,000 schizophrenia pa-tients treated with a specific type of psychiatric drug. The existing records may be used to look at improvement of say 1,000 patients who were treat-ed during 2003. This approach to a study is possible if the records on follow-ups are complete and adequately detailed and if the investigator can ascertain the current state of the patients. Schematic representation of historical control trial design is given in Figure 3.3.

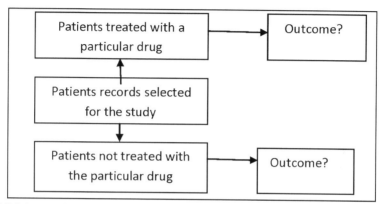

FIGURE 3.3 Schematic representation of historical control study design.

Some investigators call this type of study a historical or retrospective cohort study because historical information is used. That is, events being evaluated actually occurred before the event of the study. Studies that merely describe an investigator's experience with a group of patients and attempt to identify features associated with a good or bad outcome fall under this category.

3.4   INFORMAL DESIGNS OF EXPERIMENTS

The informal designs of experiments are those that normally use a less sophisticated form of analysis to measure the effect of treatment.

3.4.1   BEFORE-AND-AFTER WITHOUT CONTROL GROUP

In this design, a single test group is selected and the dependent variable (DV) is measured before the introduction of the treatment. The treatment is then introduced, and the dependent variable is measured again after the treatment has been introduced. Schematic representation of before-and-after without control group is given in Figure 3.4.

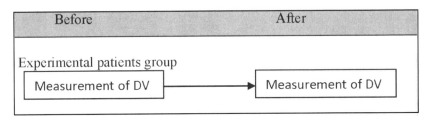

FIGURE 3.4   Schematic representation of before-and-after without control group design.

The effect of the treatment would be equal to the level of the phenomenon after the treatment minus the level of the phenomenon before the treatment. The main difficulty here is that with the passage of time, considerable extraneous variation may be there in the treatment effect.

3.4.2   AFTER-ONLY WITH CONTROL GROUP

In this experiment design, two groups, experimental group and control group, of patients are selected and the treatment is introduced into the experimental group only. The dependent variable is then measured in both the groups at the same time. Here, the two groups are identical with respect to their behavior toward the phenomenon considered. Schematic representation of after-only with control group is given in Figure 3.5.

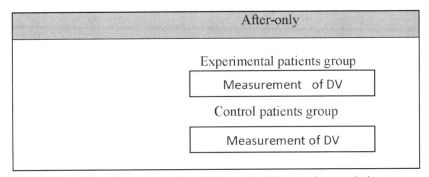

FIGURE 3.5   Schematic representation of after-only with control group design.

The treatment impact is assessed by subtracting the value of the dependent variable in the control group from its value in the experimental group. The advantage of this design is that the data can be collected without introducing problem with the passage of time.

3.4.3 BEFORE-AND-AFTER WITH CONTROL GROUP

In this design, two groups are selected, and the dependent variable is measured in both the groups for an identical time period before the treatment. The treatment is then introduced into the experimental group only, and the dependent variable is measured in both for an identical time period after the introduction of the treatment. Schematic representation of before-and-after with control group is given in Figure 3.6.

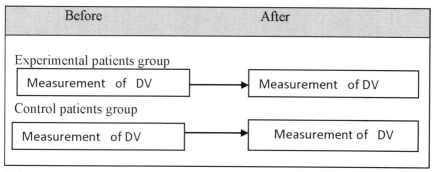

FIGURE 3.6   Schematic representation of before-and-after with control group design.

The treatment effect is determined by subtracting the change in the dependent variable in the control group from the change in the dependent variable in the experimental group. This design avoids extraneous variation resulting both from the passage of time and from noncomparability of the experimental and control groups.

3.5   FORMAL DESIGNS OF EXPERIMENTS

3.5.1   COMPLETELY RANDOMIZED DESIGN

The completely randomized design (CRD) is the simplest of all the experimental designs based on the principles of replication and randomization. Let us suppose that there are k treatments and the ith treatment is being replicated n_i times. In CRD, the whole of the experimental units are distributed completely at random to the treatments subject to the condition that the ith treatment occurs n_i times. Randomization assumes that extraneous factors do not continuously influence one treatment.

Layout: The layout of CRD is as given below:

Treatment 1	Treatment 2	-	Treatment k
x_{11}	x_{21}		x_{k1}
x_{12}	x_{22}		x_{k2}
-	-		-
x_{1n_1}	x_{2n_2}	-	x_{kn_k}

If some of the observations for any treatment are lost, then the standard analysis can be carried out on the available data. The analysis of the data in CRD is analogous to the ANOVA for one-way classification.

3.5.2   RANDOMIZED BLOCK DESIGN

If the whole of the experimental material is not homogenous, then the randomized block design (RBD) is the simple method of controlling the variability of the experimental material. The RBD consists in grouping the experimental material into relatively homogenous strata called blocks, and the treatments are applied at random to the units within each block and replicated over all the blocks. The treatments are allocated at random within the units of each block, and thus randomization is restricted. Thus, if it is desired to control one source of variation by stratification, the experimenter should select RBD rather than CRD.

Layout: The layout of RBD is as given below:

Blocks	Treatment 1	Treatment 2	-	Treatment k
1	x_{11}	x_{21}		x_{k1}
2	x_{12}	x_{22}		x_{k2}
-	.	.		.
r	x_{1r}	x_{2r}	-	x_{kr}

The statistical analysis of data in RBD is analogous to two-way ANOVA.

3.5.3 FACTORIAL DESIGNS

In factorial experiments, the effects of several factors of variation are studied and investigated simultaneously. Here, the treatments are the combinations of different factors under study. In these experiments, an attempt is made to estimate the effects of each of the factors and also their interactive effects. Let us suppose that there are p different doses of diazepam and q different doses of nitrazepam. The p and q are termed as the levels of the factors diazepam and nitrazepam, respectively. A series of experiments in which only one factor is varied at a time would be lengthy, costly, and unsatisfactory because of systematic change in the general background conditions. Moreover, these simple experiments do not tell us anything about the interaction effect. Alternatively, we try to investigate the variation in several factors simultaneously by conducting a $p \times q$ factorial experiment. In general, if the levels of various factors are equal, then r^s experiment means a factorial experiment with s factors each at r levels.

3.5.3.1 2^2-FACTORIAL DESIGN

In 2^2 (or 2×2) factorial experiment, we have two factors each at two levels, and hence there are four treatment combinations in all. Let the capital letters A and B indicate the name of the two factors under study. Let the small letters "a" and "b" denote one of the two levels of each of the corresponding factors, and this will be usually called the second level. The first level of A and B are generally expressed by the absence of the corresponding small letter in the treatment combinations. The four treatment combinations may be enumerated as follows:

1: Both A and B are at the first level

a: A is at second level and B is at the first level
b: A is at first level and B is at the second level
ab: Both A and B are at the second level

Layout: The layout of 2^2-factorial design with RBD is given below:

Blocks	Treatments			
	1	**a**	**b**	**ab**
1	x_{11}	x_{21}	x_{31}	x_{41}
2	x_{12}	x_{22}	x_{32}	x_{42}
-				
r	x_{1r}	x_{2r}	x_{3r}	x_{4r}
Total	(1)	(a)	(b)	(ab)

3.5.4 ANALYSIS OF COVARIANCE DESIGN

In analysis of variance (ANOVA), it is assumed that the observed values (y) are attributed to the treatments applied and not to any other causal circumstances. Sometimes, some concomitant variable (x) is correlated with the dependent variable (y). The marks scored in psychiatry by a group of multipurpose health workers after conducting a training program may be related to the marks scored by the group before conducting the training program. In such situations, the researcher should use the statistical technique of ANCOVA for valid comparison of treatment effects.

Layout: The layout of the ANCOVA design is as given below:

Treatment 1		Treatment 2			Treatment k	
x	y	x	y	-	x	y
x_{11}	y_{11}	x_{21}	y_{21}		x_{k1}	y_{k1}
x_{12}	y_{12}	x_{22}	y_{22}		x_{k2}	y_{k2}
.	.	.	.	.	.	.
x_{1n_1}	y_{1n_1}	x_{2n_2}	y_{2n_2}	.	x_{kn_k}	y_{kn_k}

3.5.5   REPEATED MEASURES DESIGN

The repeated measures design (also known as within-subject design) uses the same subjects with every condition of the research. For instance, repeated measures are collected in a longitudinal study in which change over time is assessed.

Layout: The layout of the design is as given below:

Subject	Repeat 1	Repeat 2	-	Repeat k
1	x_{11}	x_{21}		x_{k1}
2	x_{12}	x_{22}		x_{k2}
-	-	-	-	-
n	x_{1n}	x_{2n}		x_{kn}

One disadvantage to the repeated measures design is that it may not be possible for each participant to be present in all conditions of the experiment, especially severely diseased subjects tend to drop out of a longitudinal study. Removing these subjects would bias the design.

KEYWORDS

- **ANCOVA**
- **Clinical trials**
- **Experimental errors**

CHAPTER 4

ONE-VARIABLE DESCRIPTIVE STATISTICS

CONTENTS

4.1   CLASSIFICATION AND TABULATION OF DATA

The observations on each variable have to be classified and presented in the form of tables. The main objectives of classification are to simplify the complex nature of the data, tailoring it up, and make it easy to grasp. The classification by person may be made according to qualitative and quantitative aspects.

4.1.1   QUALITATIVE VARIABLE DATA

In the classification of a qualitative variable data, a natural classification system may be available. For example, in the classification of patients according to sex, it is enough if we count the number of males and the number of females. The classified data have to be presented in the form of tables separately for each variable or for groups of variables. The total of the frequencies and the relative frequencies in the table will aid in the interpretation of the results. Appropriate title must be given at the top of the table. The source of data and the new terms used must be specified at the bottom of the table. For example, the sex distribution of the 40 registered psychiatric patients (data in Appendix I) is as given in Table 4.1.

TABLE 4.1   Sex distribution of 40 psychiatric patients

Sex	Frequency
Males	24
Females	16
Total	40

4.1.2   QUANTITATIVE VARIABLE DATA

4.1.2.1   ARRAY

The first step in the classification of a quantitative variable data is to arrange the raw data in an array of ascending or descending order.

Example: The array of the 40 registered patients (data in Appendix I) according to age is as given below:

4	9	10	11	12	12	13	16	19	21
22	23	24	24	25	25	25	26	26	27
27	29	29	30	31	33	35	35	40	44
48	50	51	58	60	60	65	68	76	80

4.1.2.2 CLASSIFICATION

In the classification of the quantitative variable data, the optimum number and length of frequency classes depend on the size, nature of the data, and the purpose of the study. Generally, the frequency classes are of equal size, mutually exclusive, and exhaustive. The main purpose of classification of quantitative variable data is to determine the nature of the frequency distribution: normal, uniform, exponential, etc.

Example: The age distribution of the 40 registered patients is as given in Table 4.2.

TABLE 4.2   Age distribution of 40 psychiatric patients

Age	Frequencies
Below 10	2
10–19	7
20–29	14
30–39	5
40–49	3
50–59	3
60–69	3
70–79	2
80–89	1
Total	40

This age distribution follows approximate normal distribution. The frequency distribution of age increases up to 20–29 age group and decreases thereafter.

4.2   GRAPHICAL REPRESENTATION OF CLASSIFIED DATA

In more complex situation of presentation, further to see the shape and pattern of frequency distribution, graphs and diagrams may be used with or without the frequency tables. They are good visual and mental aids for both professionals and laymen. Generally, the graphs are drawn for quantitative variable data and diagrams are drawn for qualitative variable data.

4.2.1   QUALITATIVE VARIABLE DATA

4.2.1.1   BAR DIAGRAM

In bar diagram, on a common base line, vertical or horizontal bars are drawn with their heights proportional to the frequencies. The bars are of equal width and successive bars are drawn at equal distances. Composite, component, or percentage bar diagrams may be drawn depending on the size and nature of the classified data. This diagram is especially preferred for discrete data.

Example: The bar diagram showing the age distributions of 10 years' duration of the 40 psychiatric patients of NIMHANS hospital (data in Appendix I) is shown as bar diagram (Fig. 4.1).

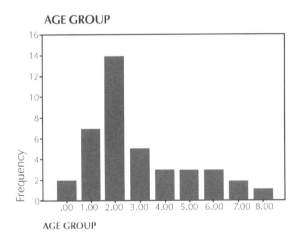

FIGURE 4.1   Bar diagram showing the age distribution of 40 psychiatric patients.

4.2.2   QUANTITATIVE VARIABLE DATA

4.2.2.1   FREQUENCY POLYGON

In frequency polygon, the frequencies are plotted on a graph sheet, and the lines joining these points are drawn. More than one frequency polygon may be drawn on the same graph sheet, and thus several frequency distributions may be compared. This graphical representation is more appropriate for continuous data as shown in Figure 4.2. It can be viewed that the psychiatric patients of young age groups and old-age groups in the community are under-represented at NIMHANS hospital.

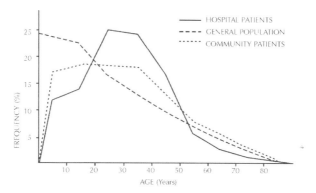

FIGURE 4.2   Frequency polygons showing age distributions of 500 psychiatric patients.

4.2.2.2   HISTOGRAM

In histogram, on a common base line, rectangles are drawn with their areas proportional to the frequencies. These rectangles are in juxtaposition, and there will be no gaps between them. This diagram is more appropriate for data in ratio scale of measurement and especially when the frequency classes are of unequal widths as shown in Figure 4.3. In order to draw this diagram, the frequencies of the classes of varying duration of illness had to be converted into frequencies for unit duration of illness.

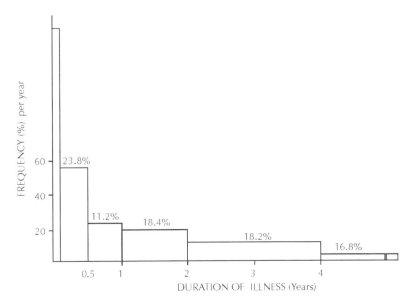

FIGURE 4.3   Histogram showing the duration of illness of 500 psychiatric patients.

4.3   SUMMARIZING DATA: AVERAGES

The summarization figures of measurement of characteristics of frequency distribution will be helpful to understand the data more clearly. The observed values of a psychiatric variable are not equal, but we can notice a general tendency of these values to cluster around a particular value. It is convenient to characterize and represent each group of observations by such a value that is called the central tendency or the average of that group.

4.3.1   QUANTITATIVE VARIABLE DATA

4.3.1.1   MEAN

The mean (arithmetic mean) is obtained by dividing the sum of observations by the total number of observations. Therefore,

$$\text{Mean} = \frac{\text{Sum of observations}}{\text{Total number of observations}}$$

Symbolically,

$$\bar{x} = \frac{1}{n} \sum x_i$$

where x_i are the observed values and n is the number of observations.

The population mean is denoted by μ, and the sample mean is denoted by $\bar{x}$. The mean is sensitive to the presence of extreme values in the data, and hence it does not give fair idea of central tendency when the extreme values or outliers are present in the data.

4.3.1.2 MEDIAN

The median is the middle most value of the observations arranged in an ascending or descending order of magnitude. It divides the distribution into two equal parts, and hence it is a position value.

4.3.1.3 MODE

The mode is that value of the observations, which occurs most frequently. There may be more than one mode in a frequency distribution.

Example: Let us suppose that the marks scored in a test (out of 10) by a group of five students are given by 7, 4, 1, 3, and 5. For this data,

Array:	1, 3, 4, 5, 7
Sample size (n):	5
Sum of observations: ($\sum x$)	20
Mean ($\bar{x}$):	4.0
Median:	4
Mode:	not available

4.3.2   QUALITATIVE VARIABLE DATA

4.3.2.1   PROPORTION AND PERCENTAGE

A proportion of a category of a qualitative variable is given by
$$P = \text{numerator/denominator}$$

A percentage is defined as follows
$$\text{Percentage, (\%)} = \text{proportion} \times 100$$
$$= p \times 100$$

Thus, the percentages of frequencies of a categorical variable (qualitative variable) add to 100, and the proportions of the frequencies add to 1.

Example: The proportions and percentages of males and females of the 40 registered patients (data in Appendix I) are given in the following table:

Sex	Frequency	Proportion	Percentage
Male	24	0.6	60.0
Female	16	0.4	40.0
Total	40	1.0	100.0

4.3.2.2   RATE AND RATIO

A rate measures diseases, disabilities, and injuries per unit population. That is,

$$\text{Rate} = \frac{\text{Numerator} \times \text{duration of time} \times \text{k}}{\text{Denominator}}$$
$$= \frac{\text{Number infected} \times \text{duration of time} \times \text{k}}{\text{Population at risk}}$$

When the denominator consists of all those who have the chance of getting the disease, it is known as the population at risk.

Example: There are 11 admissions for in-patients services out of a total of 40 registered patients. Then,

$$\text{Rate of admission} = \frac{11 \times 100}{40} = 27.5\%$$

The rates may be crude (crude birth rate), specific (sex specific rate), or standardized (standardized death rate).

A ratio is defined as follows:

$$\text{Ratio} = \frac{\text{Numerator} \times \text{point of time} \times k}{\text{Denominator}}$$

$$= \frac{\text{Number infected} \times \text{point of time} \times 1000}{\text{Population at risk}}$$

Thus, in a ratio, the numerator and the denominator are independent. A rate refers to a period of time, whereas a ratio refers to a point of time. For examples, the sex ratio of Indian population is 940 females per 1,000 males.

4.3.2.3   INDICATORS AND INDEX

An indicator can be defined as something that helps us understand what we are, where we are going, and how far we are from the goal. Therefore, it can be a number or a graphic. Health indicators are established to monitor the health status of a given population and to evaluate the number of individuals affected by a particular disease within a given time period and the total number of cases of that disease. It is also used to evaluate health risks and performance of health system. An index is a statistic giving the value of a quantity such as cost per day per patient, etc., relative to its level at some fixed time or place, which conventionally is given the number hundred.

4.4   SUMMARIZING DATA: DISPERSION

The observed values of a variable tend to spread over an interval rather than cluster closely around the central average. The characteristics of the scatter or spread of the observed values in the neighborhood of the central average is called dispersion.

4.4.1   QUANTITATIVE VARIABLE DATA

4.4.1.1   RANGE

The minimum value, the maximum value, and the range of observed values of a variable are the crude measures of dispersion. The range is given by

$$R = X_{maximum} - X_{minimum}$$

4.4.1.2   STANDARD DEVIATION

The range depends only on the extreme values, and it does not deal with the variation within the group. The standard deviation is the root-mean-squared deviation measured from the mean. The population standard deviation is denoted by σ and that of the sample by S (or SD). Thus,

$$S = \sqrt{\frac{1}{n}\sum\left(x - \bar{x}\right)^2}$$

4.4.1.3   COEFFICIENT OF VARIATION

The coefficient of variation (CV) is a relative measure of dispersion. It is given by the following relation:

$$CV = \frac{Standard\ deviation \times 100}{Mean}$$

Example: Let us suppose that the marks scored in a test by a group of five students are given by 7, 4, 1, 3, and 5. For this data,

Minimum value = 1
Maximum value = 7
Range = 7 − 1 = 6
To calculate standard deviation, the computation table is given below:

Student	x	$(x - \bar{x})$	$(x - \bar{x})^2$
1	7	3	9
2	4	0	0
3	1	−3	9
4	3	−1	1
5	5	1	1
Total	20	0	20
Mean	4.0	0	-

Now, the standard deviation is calculated as follows:

$$S = \sqrt{\frac{20}{5}} = \sqrt{4} = 2.0$$

The coefficient of variation is calculated as follows:

$$CV = \frac{2 \times 100}{4} = 50\%$$

4.4.2   QUALITATIVE VARIABLE DATA

4.4.2.1   FREQUENCY DISTRIBUTION

The frequency distribution and the relative frequency distribution serve as the measurement of dispersion in case of qualitative variable data. For example, the diagnostic distribution of the 40 registered patients (data in Appendix I) is as given in Table 4.3.

TABLE 4.3   Diagnostic distribution of 40 psychiatric patients

Disorder	Number of Patients	(%)
1. Organic psychosis	2	5.0
2. Substance use disorder	1	2.5
3. Schizophrenia	8	20.0
4. Mania	-	-
5. Bipolar affective disorder	8	20.0
6. Endogenous depression	7	17.5
7. Phobia	-	-
8. Obsessive compulsive neuroses	2	5.0
9. Anxiety neurosis	1	2.5
10. Neurotic depression	2	5.0
11. Dissociative disorder	2	5.0
12. Somatoform disorder	-	-
13. Behavioral syndrome	-	-
14. Personality disorder	-	-
15. Mental retardation	2	5.0
16. Developmental disorder	4	10.0
17. Behavior/emotional disorder	1	2.5
18. Epilepsy	-	-
All disorders	40	100.0

4.5 SUMMARIZING DATA: SKEWNESS AND KURTOSIS

4.5.1 SKEWNESS

The skewness is the lack of symmetry. A perfectly bell-shaped curve has no skewness. Hence, a distribution is skew if the mean, median, and mode do not have the same value. The skewness may be positive such as age distribution of dissociative disorder patients (CON), negative such as age distribution of somatoform (SOM) disorder patients, or zero such as age distribution of obsessive and compulsive disorder (OCD) patients at NIM-HANS hospital as shown in Figure 4.4.

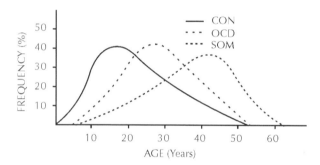

FIGURE 4.4 Frequency curves showing positively skewed, negatively skewed, and symmetrical distributions.

4.5.1.1 COEFFICIENT OF SKEWNESS

One of the measures of skewness is coefficient of skewness that usually lies between −1 and 1. It is given by

$$\text{Coefficient of skewness} = \frac{\text{Mean} - \text{Mode}}{\text{Standard deviation}}$$

4.5.2 KURTOSIS

Kurtosis is the measure of the relative flatness of the top of the frequency curve. A frequency curve may be symmetrical, but it may fail to have peakedness as that of a normal curve. The frequency distribution may have the same variability, but this may be more or less peaked than that of the normal curve. The kurtosis may be classified as mesokurtic (medium kurtosis), leptokurtic (high kurtosis), and platykurtic (low kurtosis) as shown in Figure 4.5, where obsessive thought disorder (OTD), obsessive compulsive disorders (OCD), dissociative disorder and somatoform disorder (DSD) at NIMHANS hospital.

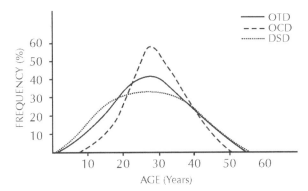

FIGURE 4.5   Frequency curves showing leptokurtic, mesokurtic, and platykurtic distributions.

4.6   DESCRIPTIVE STATISTICS FOR SPATIAL DATA

4.6.1   CLASSIFICATION OF SPATIAL DATA

The geographical data are sometimes arranged alphabetically. The spatial characteristic of the data when presented in tables is largely obscured. It is advantageous to incorporate both the geographical and political aspects in maps.

4.6.2   SPOT MAPS

The maps provide the necessary medium for presenting real relationship of spatial data clearly, meaningfully, and adequately. Maps are often indispensible in locating problems, analyzing data, and discovering hidden facts and relationships. It is advantageous to incorporate both the geographical and political aspects in maps. The geographical aspect is important to determine the causal factors of disorders/diseases. The political aspect is important for effective mental health delivery systems.

There are spot maps, shaded maps, and superimposed maps. In spot maps, the size of the symbol may be proportional to the magnitude of the phenomena represented. Symbols may be in the form of circles. The area of the circle may be proportional to the value represented.

Example: The 36 government mental hospitals in India are presented as spot map as shown in Figure 4.6.

FIGURE 4.6 Spot map portraying bed occupancies in government mental hospitals in India.

The geographical area that is too far to a government mental hospital in India may be located using the spot map.

4.6.3 DENSITY OF PATIENTS

It is given by

$$\text{Density of patients} = \frac{\text{Total number of patients}}{\text{Area in square kilometers}}$$

Example: The density of psychiatric patients in India is calculated as follows:

$$\text{Density of patients} = \frac{7,20,00000}{33,00000} = 21.8 \text{ per square kilometer}$$

4.7 DESCRIPTIVE STATISTICS FOR TIME-RELATED DATA

4.7.1 CLASSIFICATION OF TIME-RELATED DATA

The classification of time period depends on the purpose of the epidemiological study and the time series analysis required for the purpose. The classification can be made according to hours, days, weeks, months, years, etc., depending on the purpose of the study.

4.7.2 EPIDEMIOLOGICAL CURVES

An epidemic curve is defined as a plot of the number of cases against time of onset of disease, with time on the horizontal axis, and the number of new cases on the vertical axis. It is a method of visualizing the progression of a disease over time, which helps epidemiologists to answer several important questions.

4.7.3 EXPECTATION OF DURATION OF TIME AND TIME SERIES COMPONENTS

The expectation of duration of time (Life tables) and Time series components (Trend, seasonal variation, cyclic components) are important statistical methods relating to time (Chapter 14).

KEYWORDS

- **Coefficient of variation**
- **Descriptive statistics**
- **Quantitative/qualitative variable data**

CHAPTER 5

MENTAL HEALTH STATISTICS

CONTENTS

5.1 DEMOGRAPHIC INDICATORS

Demography is the study of the composition, distribution, and growth of human population, and their interrelationships with social, economic, and behavioral factors. In addition to statistics relating to existing mental health conditions and available mental health care facilities, knowledge of demographic indicators is important for psychiatric service and research for specification of goals and targets in terms of measurable outputs.

5.1.1 POPULATION SIZE

Population size is an important figure as it is utilized in deriving various mental morbidity indices.

5.1.1.1 MALTHUS THEORY

This theory states that the human population will expand exponentially, whereas the means of subsistence will increase more slowly, with the consequence that ultimately the size will be kept in check by famine, disease (including mental disorders), and war.

5.1.2 POPULATION COMPOSITION

5.1.2.1 AGE DISTRIBUTION

Age distribution of the population provides the true picture regarding the risk of various mental and behavioral disorders associated with different age groups.

5.1.2.2 SEX DISTRIBUTION

Sex distribution of the population will help in studying different mental and behavioral disorders associated with each sex. This is important for improving the mental health care of the entire population.

5.1.2.3   DOMICILE DISTRIBUTION

The domicile (rural–urban) distribution of the population will help in preparing the required mental health programs for these sectors separately.

5.1.2.4   RELIGION/CASTE DISTRIBUTIONS

Most of the psychiatric disorders are associated with religion/caste distributions.

5.1.2.5   FAMILY SIZE

The average family size is also important since it varies with several psychiatric disorders.

5.1.3   POPULATION DISTRIBUTION

The population distribution of a geographical region can be studied by calculating the density of population.

5.1.3.1   DENSITY OF POPULATION

The density of population per square kilometer of a given geographical region is given by the following relation:

$$\text{Density of population} = \frac{\text{Total population of the area}}{\text{Area in square kilometers}}$$

The density of population in urban locality may be calculated as per persons per room to determine overcrowding.

5.1.4   POPULATION GROWTH

The growth rate of population in a given geographical region is studied by calculating the growth rate of population.

5.1.4.1   ANNUAL GROWTH RATE OF POPULATION

It is given by the following relation:

$$\text{Annual growth rate} = \frac{\text{Increase /decrease in population during the year} \times 1000}{\text{Population as on 1 July of the year}}$$

5.1.5   MEASURES OF FERTILITY AND MORTALITY

A knowledge of the rates of fertility and mortality in a community is important for planning mental health facilities and services for various groups of people.

5.1.5.1   CRUDE BIRTH RATE

The annual crude birth rate (CBR) is the general rate of fertility. It is defined as follows:

$$\text{CBR} = \frac{\text{Number of live births during the year} \times 1000}{\text{Population as on July 1 of the year}}$$

5.1.5.2   CRUDE DEATH RATE

The annual crude death rate (CDR) is the general rate of mortality. It is defined as follows:

$$\text{CDR} = \frac{\text{Number of deaths during the year} \times 1000}{\text{Population as on July 1 of the year}}$$

5.2   DEMOGRAPHIC INDICATORS OF INDIA

The demographic indicators of India for the year 2012 are in Table 5.1.

TABLE 5.1   Demographic indicators of India

1. Population size (in crores)		123
2. Age distribution (%):	0–9	19.6
	10–19	18.7
	20–29	19.4
	30–39	14.4
	40–49	11.6
	50–59	8.1
	60–69	5.0
	70–79	2.4
	80–	0.8
3. Sex distribution (%):	Male	51.5
	Female	48.5
4. Domicile distributions (%):	Rural	68.8
	Urban	31.2
5. Religion distribution (%):	Hindu	80.5
	Muslim	13.4
	Christian	2.3
	All others	3.8
6. Average family size		4.8
7. Density of population (per square kilometer)		382
8. Annual growth rate (%)		1.4
9. Crude birth rate		20.6
10. Crude death rate		7.4

5.3   MEASURES OF MENTAL MORBIDITY

The prevalence rates and the incidence rates are the important indices to measure epidemiological situation, suggest priorities, and assess the progress made in the control of diseases/disorders.

5.3.1 PREVALENCE RATE

The prevalence rate (point) is the rate at which people having the disease at a particular point of time. That is,

$$\text{Prevalence rate} = \frac{\text{Number of cases at a particular point of time} \times 1000}{\text{Population at risk at the particular point of time}}$$

The prevalence rates are determined for chronic illnesses such as mental and behavioral disorders. The lack of absolute standard criteria for defining a case is the foremost among the methodological problems in psychiatric epidemiological field.

5.3.2 INCIDENCE RATE

The incidence rate is the rate at which people without the disease develop the disease during a specified period of time (say 1 year). That is,

$$\text{Incidence rate} = \frac{\text{Number of new cases during the period} \times 1000}{\text{Population at risk during the period}}$$

The incidence rates are usually determined for acute diseases such as infectious diseases, as their onsets are sharply defined. However, in community psychiatry, these rates are also important to answer the question of whether the growth of urbanization leads to increase in mental stress and hence mental disorders.

5.4 MENTAL MORBIDITY IN INDIA

5.4.1 PREVALENCE RATES AND INCIDENCE RATES

The prevalence rate of mental and behavioral disorders in India is estimated to be 58.2 per 1,000 population. The annual incidence rate of mental and behavioral disorders in India is estimated to be 17.1 per 1,000 population. The details are given in Table 5.2.

TABLE 5.2 Prevalence rates and incidence rates of mental and behavior disorders

Disorders	Prevalence Rate	Annual Incidence Rate
1. Organic psychosis	0.4	0.6
2. Substance use disorder	6.9	0.1
3. Schizophrenia	2.7	1.2
4. Mania	0.7	0.1
5. Bipolar affective disorder	2.7	0.9
6. Endogenous depression	8.9	5.4
7. Phobia	4.2	1.8
8. Obsessive compulsive disorder	3.1	0.3
9. Anxiety neurosis	5.8	1.8
10. Neurotic depression	3.1	1.2
11. Dissociative disorder	4.5	1.5
12. Somatoform disorder	0.6	0.9
13. Behavioral syndrome	0.1	0.1
14. Personality disorder	0.5	0.1
15. Mental retardation	6.8	0.3
16. Developmental disorder	0.1	0.1
17. Behavioral/emotional disorder	2.7	0.6
18. Epilepsy	4.4	0.1
All disorders	58.2	17.1

5.4.2 AGE-WISE PREVALENCE RATES

Age-wise prevalence rates of mental and behavioral disorders in India are given in Table 5.3

TABLE 5.3 Age-wise prevalence rates of mental and behavioral disorders in India (rate per 1,000 population)

	Disorder	Combined	0–9	10–19	20–29	30–39	40–49	50–59	60–69	70–79	80–
1	Organic psychosis	0.4	–	–	–	0.2	0.6	1.3	2.0	2.6	**4.1**
2	Substance use disorder	6.9	–	1.9	4.5	8.1	15.9	**20.6**	15.4	6.3	–
3	Schizophrenia	2.7	–	1.1	4.3	**5.6**	5.2	3.6	1.1	–	–
4	Mania	0.7	–	0.2	1.0	**2.0**	1.7	0.5	–	–	–
5	Bipolar affective disorder	2.7	–	–	2.5	6.1	**7.4**	4.2	3.0	2.1	1.9
6	Endogenous depression	8.9	–	0.4	4.9	11.7	17.2	21.7	27.7	32.0	**33.7**
7	Phobic neurosis	4.2	3.7	**12.7**	5.9	–	–	–	–	–	–
8	Anxiety neurosis	5.8	2.2	6.4	7.5	**8.9**	6.0	4.4	3.0	0.6	–
9	Obsessive compulsive neurosis	3.1	–	1.3	**7.0**	6.5	4.2	0.9	–	–	–
10	Neurotic depression	3.1	–	1.4	3.6	5.1	**6.2**	5.2	4.8	3.2	–
11	Dissociative disorder	4.5	–	0.1	5.5	**10.8**	9.7	8.0	2.0	–	–
12	Somatoform disorder	0.6	–	0.1	0.3	0.8	1.6	**1.8**	1.3	1.0	0.6
13	Behavioral syndrome	0.1	–	–	0.1	**0.3**	0.1	–	–	–	–
14	Personality disorder	0.5	–	–	0.7	**2.0**	0.7	–	–	–	–
15	Mental retardation	6.8	**14.4**	12.9	5.8	2.7	0.4	–	–	–	–
16	Developmental disorder	0.1	**0.3**	0.2	–	–	–	–	–	–	–
17	Behavioral/ emotional disorder	2.7	3.3	**7.4**	3.4	–	–	–	–	–	–
18	Epilepsy	4.4	4.8	**7.1**	5.7	5.0	2.1	0.4	–	–	–
	All disorders	58.2	28.7	53.2	62.7	75.8	**79.0**	72.6	60.3	47.8	40.3

5.4.2.1 REMARKS

It can be noted that in psychiatric patient population, the prevalence rates increase with age exponentially in respect of organic psychosis and endogenous depression. The prevalence rates are highest in the youngest age groups and decrease gradually in respect of mental retardation and developmental disorders. In the remaining diagnostic categories, the distributions are asymptotic normal.

5.4.3 SEX-WISE PREVALENCE RATES

Sex-wise prevalence rates of mental and behavioral disorders in India is given in Table 5.4.

TABLE 5.4   Sex-wise prevalence rates of mental and behavioral disorders in India

Disorders	Combined	Male	Female
1. Organic psychosis	0.4	0.2	0.6
2. Substance use disorder	6.9	11.9	1.7
3. Schizophrenia	2.7	2.3	3.2
4. Mania	0.7	1.2	0.1
5. Bipolar affective disorder	2.7	2.3	3.2
6. Endogenous depression	8.9	5.6	12.3
7. Phobia	4.2	2.4	6.0
8. Obsessive compulsive disorder	3.1	1.7	4.6
9. Anxiety neurosis	5.8	3.2	8.4
10. Neurotic depression	3.1	1.2	5.2
11. Dissociative disorder	4.5	1.2	8.0
12. Somatoform disorder	0.6	0.7	0.4
13. Behavioral syndrome	0.1	0.1	0.1
14. Personality disorder	0.5	1.0	0.2
15. Mental retardation	6.8	9.1	4.4
16. Developmental disorder	0.1	0.1	0.1
17. Behavioral/emotional disorder	2.7	2.7	2.7
18. Epilepsy	4.4	4.9	3.9
All disorders	58.2	51.9	64.8

5.4.3.1 REMARKS

The prevalence rate was 51.9 for males and 64.8 for females. Manic affective psychosis, mental retardation, alcohol/drug addiction, and personality disorders were significantly high in males, whereas organic psychosis, endogenous depression, and all neurotic disorders were significantly high among females.

5.4.4 DOMICILE-WISE PREVALENCE RATES

Domicile wise-prevalence rates of mental and behavioral disorders in India are given in Table 5.5

TABLE 5.5 Domicile wise-prevalence rates of mental and behavioral disorders in India

Disorders	Combined	Rural	Urban
1. Organic psychosis	0.4	0.6	0.1
2. Substance use disorder	6.9	7.3	5.8
3. Schizophrenia	2.7	2.6	2.9
4. Mania	0.7	0.6	0.8
5. Bipolar affective disorder	2.7	2.7	3.0
6. Endogenous depression	8.9	7.7	14.1
7. Phobia	4.2	2.4	8.0
8. Obsessive compulsive disorder	3.1	2.3	5.0
9. Anxiety neurosis	5.8	3.1	11.6
10. Neurotic depression	3.1	1.1	7.7
11. Dissociative disorder	4.5	5.0	3.4
12. Somatoform disorder	0.6	0.8	0.2
13. Behavioral syndrome	0.1	0.1	0.1
14. Personality disorder	0.5	0.5	0.7
15. Mental retardation	6.8	6.3	8.8
16. Developmental disorder	0.1	0.1	0.1
17. Behavioral/emotional disorder	2.7	1.0	5.9
18. Epilepsy	4.4	4.8	2.5
All disorders	58.2	48.9	80.6

5.4.4.1   REMARKS

The prevalence rate was 48.9 for rural sectors and 80.6 for urban sectors. Only epilepsy and hysteria were significantly high in rural communities, whereas endogenous depression, mental retardation, all neurotic disorders (except hysteria), and behavioral/emotional disorders were significantly high in urban communities.

5.5   MENTAL HEALTH DELIVERY SYSTEMS IN INDIA

5.5.1   CLASSIFICATION

A broad classification of mental health delivery systems in India is as given below:
1. Mental Health Institutes (Hospitals)
    (a) State government mental hospitals
    (b) Central government mental hospitals
    (c) Autonomous body hospitals
    (d) Private hospitals
2. Psychiatric Rehabilitation Centers
3. Hospital Psychiatric Units
    (a) Medical college hospital psychiatric units
    (b) District hospital psychiatric units
    (c) Specialized hospital psychiatric units
4. Psychiatry Clinics
5. Prison Psychiatric Services
6. Community Mental Health Delivery Systems
    (a) Satellite clinics
    (b) Mental morbidity studies
    (c) Indian system of medicine

5.6   MENTAL HEALTH SERVICE INDICATORS

5.6.1   OUT-PATIENT SERVICE INDICATORS (YEAR)

The out-patients service indicators are common for all mental health delivery systems irrespective of whether in-patients service is available or not. These indicators refer to a particular time period usually 1 year, and the data collected are subjected to case-series analysis. The out-patients population forms two sub-populations: newly registered patients and the follow-up patients. The important indicators are

1. Number of new registrations (year)
2. Age distribution
3. Sex distribution
4. Diagnostic distribution
5. Rate of follow-up attendance

$$\text{Rate of follow-up attendance} = \frac{\text{Total follow-up attendance during the year}}{\text{Number of new registrations during the year}}$$

The characteristics of the registered patients are based on the disorders/diseases for which the mental health facilities are being sought, and hence such data may not be a fair representation of the diseases/disorders in the surrounding community. Still they have several administrative and clinical values. Although these data do not provide the estimation at the community level, they are used in spelling out the dimensions of the problems and their applications in planning process and quality-of-life ascertainment. The annual seasonal variation of the disorders in the community may be reflected by changes in the new registrations of the hospital.

5.6.2 ADMITTED PATIENTS (YEAR)

The admitted patients for in-patients treatment form a sub-population of the registered patients. Hence, it is restricted only to mental health institutes (hospitals) and hospital psychiatric units where in-patients facilities are available. Again, these indicators refer to a particular time period (say 1year). The important indicators are the following:

6. Number of admitted patients (year)
7. Rate of admission

$$\text{Rate of admission (\%)} = \frac{\text{Number of patients admitted during the year} \times 100}{\text{Number of new registrations during the year}}$$

5.6.3 IN-PATIENT SERVICE INDICATORS (JULY 1 OF THE YEAR)

The characteristic distributions of in-patients on a particular day (say July 1 of the year) is an important aspect to compare and evaluate the services of a mental health institute (hospital) and hospital psychiatric units. This can be done by conducting a census (all the in-patients in the hospital) on that day. The important indicators are the following:

8. Number of in-patients (July 1 of the year)
9. Age distribution
10. Sex distribution
11. Diagnostic distribution
12. Mode of admission status (this may be voluntary, certified, observation, or criminal records)
13. Duration of stay

5.6.4 DISCHARGED CASES (YEAR)

The important indicators are the following:

14. Number of discharged cases (year)
15. Result of treatment
16. Hospital death rate

$$\text{Hospital death rate} = \frac{\text{Number of deaths during the year in the hospital} \times 1000}{\text{Number of in-patients on July 1 of the year}}$$

The result of treatment may be classified as recovered, improved, slightly improved, not improved, and expired.

5.6.5 MAN-POWER INDICATORS (JULY 1 OF THE YEAR)

The important indicators are the following:

17. Number of psychiatrists (July 1 of the year)
18. In-patients per psychiatrist

$$\text{In-patients per psychiatrist} = \frac{\text{Number of in-patients on July 1 of the year}}{\text{Number of psychiatrists on July 1 of the year}}$$

5.6.6 EXPENDITURE PATTERN INDICATORS (FINANCIAL YEAR)

The important indicators are the following:

19. Total expenditure on service (financial year)
20. Unit cost (cost per day per patient)

$$\text{Unit cost} = \frac{\text{Total expenditure on service during the financial year}}{\text{Hospital days of the financial year}}$$

The hospital days of the year are the sum of the daily census figures of all the 365 (or 366) days of the year.

5.7   A PROFORMA TO COLLECT BASIC INFORMATION OF MENTAL HOSPITALS

Name & full address of the hospital   ....

....

....PIN   ....

Purpose   Service/Training/Research/Others(specify)

Administered by   Autonomous body/Government/Private/Others(specify)

STATISTICS – (Year ....)

1 Bed Strength		2 Hospital days	

3 Strength (in-patients) as on 1st July of the year	Less than 2 years	2-5 Years	5-10 years	More than 10 years	Total

4 Discharged Patients

a	Voluntary		Certified	Observations	Criminal	Total
	Paying	Free				

b	Psychoses		Neuroses	M.R.	Total
	Organic	Functional			

c	Improved	Not improved	Deaths	Total

5	Out-Patient Treatment	New Registration	Follow-up Attendance	Total

6	Expenditure as on :				
	Diet	Medicine	Salaries	Others (specify)	Total

5.8   SERVICE INDICATORS OF GOVERNMENT MENTAL HOSPITALS IN INDIA

There are 36 government mental hospitals in India.

5.8.1   SERVICE INDICATIONS

TABLE 5.6: Basic information and service indicators of government mental hospitals (2002)

Sl. No.	Service Indicators	Service Indicators in India	Computation
1	Number of in-patients (as on 1.7.2002)	14,579	Basic data
2	Number of registrations	1,72,909	Basic data
3	Follow-up attendance	7,85,388	Basic data
4	Rate of follow-up attendance	4.5	7,85,388/1,72,909
5	Number of admissions	51,539	Basic data
6	Rate of admissions	29.8	51,539/1,72,909
7	Rate of registrations to in-patients	11.9	1,72,909/14,579
8	Number of chronic patients	6,940	Basic data
9	Percentage of chronic patients	47.6	6,940/14,579
10	Number of deaths	474	Basic data
11	Hospital death rate (per 1000 in-patients)	32.5	474/14,579
12	Number psychiatrists	205	Basic data
13	In-patients per psychiatrist	71	14,579/205

5.8.2   OUT-PATIENT SERVICE INDICATORS

Out-patient services indicators of government mental hospitals are given in Table 5.7.

TABLE 5.7   Out-patient services indicators of government mental hospitals

Out-patient Service Indicators

1. Number of new Registrations (2002): Ranged from 27 to 9215 at individual hospitals

2. Age (%):	Below 17	15.4
	17–60	81.0
	Above 60	3.6
3. Sex (%)	Male (%):	67.1
	Female (%):	32.9
4. Diagnosis (%):		
	Organic psychoses	2.5
	Substance use disorders	6.4
	Schizophrenia	49.5
	Mania	6.6
	Bipolar affective disorders	12.5
	Endogenous depression	7.5
	Phobic neuroses	0.4
	Anxiety neuroses	1.9
	Obsessive compulsive neuroses:	0.7
	Neurotic depression	2.8
	Dissociative disorders	0.5
	Somatoform disorders	0.4
	Behavioral disorders	0.3
	Personality disorders	0.2
	Mental retardation	4.4
	Developmental disorders	0.1
	Behr./emotional disorders	0.3
	Epilepsy	3.0
5. Rate of follow-up attendance		4.5

5.8.3   IN-PATIENT SERVICE INDICATORS

In-patient services indicators of government mental hospitals are given in Table 5.8.

TABLE 5.8 In-patient services indicators of government mental hospitals

Admitted patients:

6. Number of admitted patients: Ranged from 12 to 5,149 at individual hospitals

7. Rate of admission (%):   38.0

In-patients service indicators:

8. Number of in-patients: Ranged from 19 to 2,411 at individual hospitals

9. Age (%):	Below 17	1.8
	17–60	88.7
	Above 60	9.5
10. Sex (%):	Male	53.0
	Female	47.0
11. Diagnosis (schizophrenia %):		62.1
12. Mode of admission status (%):		
	Voluntary	56.3
	Certified	37.2
	Observation	4.0
	Criminal	2.5
13. Duration of staying (%)		
	Below 2 years	52.4
	2–5 years	17.6
	5–15 years	18.1
	Above 15 years	11.9

Discharged cases:

14. Number of discharged cases (2012): Ranging from 10 to 5,257 at individual hospitals

15. Result of treatment (%):		
	Recovered	17.2
	Improved	50.8
	Slightly improved	29.1
	Not improved	2.0
	Expired	0.9
16. Hospital death rate:		33.0

5.9   COMPARISON OF PSYCHIATRIC PATIENT POPULATION, GOVERNMENT MENTAL HOSPITALS, AND GENERAL HOSPITAL PSYCHIATRY UNITS IN INDIA

5.9.1   DIAGNOSTIC DISTRIBUTIONS

TABLE 5.9   Diagnostic distribution of three groups of psychiatric patients

Sl. No.	Disorder	Psychiatric Patients Population (%)	Government Mental Hospitals (%)	General Hospital Psychiatric Units (%)
1	Organic psychosis	0.7	2.5	**4.9**
2	Substance use disorder	**11.9**	6.4	10.7
3	Schizophrenia	4.7	**49.5**	20.3
4	Mania	1.2	**6.6**	2.5
5	Bipolar affective disorder	4.6	**12.5**	4.5
6	Endogenous depression	15.3	7.5	**20.8**
7	Phobic neurosis	**7.2**	0.4	1.6
8	Anxiety neurosis	**10.0**	1.9	7.1
9	Obsessive compulsive neurosis	**5.3**	0.7	1.9
10	Neurotic depression	**5.3**	2.8	4.6
11	Dissociative disorder	**7.7**	0.5	4.0
12	Somatoform disorder	1.0	0.4	**4.5**
13	Behavioral syndrome	0.2	0.3	**2.5**
14	Personality disorder	**0.9**	0.2	0.8
15	Mental retardation	**11.7**	4.4	5.1
16	Developmental disorder	0.2	0.1	**0.7**
17	Behavioral / emotional disorder	**4.6**	0.3	1.6
18	Epilepsy	**7.5**	3.0	1.9
	All disorders	100.0	100.0	100.0

From Table 5.9, it can be noted that in the psychiatric patient population, the prevalence of substance use disorders, all neurotic disorders, personality disorders, mental retardation, behavioral and emotional disorders of children, and epilepsy are highest. In governmental mental hospitals, schizophrenia, mania, and bipolar affective disorders are highest. In general hospital psychiatry units, organic psychosis, endogenous depression, somatoform disorders, behavioral syndromes, and developmental disorders are highest.

5.9.2 AVERAGE (MEAN) AGES OF DIAGNOSTIC CATEGORIES

TABLE 5.10   Average (mean) ages of three groups of psychiatric patients

Sl. No.	Disorder	Psychiatric Patients Population (Mean Age)	Government Mental Hospitals (Mean Age)	General Hospital Psychiatric Units (Mean Age)
1	Organic psychosis	58.8	46.8	38.3
2	Substance use disorder	43.7	38.0	34.1
3	Schizophrenia	35.1	33.9	28.9
4	Mania	34.5	32.9	25.3
5	Bipolar affective disorder	40.4	35.3	33.7
6	Endogenous depression	49.1	41.8	34.9
7	Phobic neurosis	20.5	34.5	28.1
8	Anxiety neurosis	30.6	38.4	28.7
9	Obsessive compulsive neurosis	28.0	28.3	25.6
10	Neurotic depression	38.8	38.6	29.1
11	Dissociative disorder	38.2	24.5	20.8
12	Somatoform disorder	46.3	36.5	28.8
13	Behavioral syndrome	34.6	34.5	26.8
14	Personality disorder	33.4	24.5	21.9
15	Mental retardation	13.4	19.8	10.3
16	Developmental disorder	8.7	15.0	7.6
17	Behavioral / emotional disorder	14.2	14.5	9.8
18	Epilepsy	21.3	25.3	20.3
	All disorders	32.2	34.5	29.3

From Table 5.10, it can be noted that in psychiatric patient population average ages of severe mental disorders, neurotic depression, dissociative disorders, somatoform disorders, behavioral syndromes, and personality disorders are highest. In governmental mental hospitals, the average ages of phobic neurosis, anxiety neurosis, and obsessive compulsive neurosis and all childhood disorders and epilepsy are highest. In general, in hospital psychiatric units, the average age of none of the diagnostic categories are highest.

5.9.3 AVERAGE AGES OF ONSET OF PSYCHIATRIC DISORDERS

Since the average age of registered patients at general hospital psychiatric units are less than that of the mental hospitals as well as that of the psychiatric patient population in all the diagnostic categories, these average ages of all diagnostic categories can be reasonably considered as the average ages of onset of these disorders. Thus, the ascending order of average age of onset of psychiatric disorders is as given in the Table 5.11.

TABLE 5.11   Average Ages of onset of Psychiatric Patients

Sl. No.	Diagnostic Categories	Average Age of Onset
1	Developmental disorder	7.6
2	Behavioral / emotional disorder	9.8
3	Mental retardation	10.3
4	Epilepsy	20.3
5	Dissociative disorder	20.8
6	Personality disorder	21.9
7	Mania	25.3
8	Obsessive compulsive neurosis	25.6
9	Behavioral syndrome	26.8
10	Phobic neurosis	28.1
11	Anxiety neurosis	28.7

TABLE 5.11   *(Continued)*

12	Somatoform disorder	28.8
13	Schizophrenia	28.9
14	Neurotic depression	29.1
15	Bipolar affective disorder	33.7
16	Substance use disorder	34.1
17	Endogenous depression	34.9
18	Organic psychosis	38.3
All disorders		29.3

5.9.4 SEX-DIAGNOSTIC-WISE DISTRIBUTION OF THREE GROUPS OF PSYCHIATRIC PATIENTS

TABLE 5.12   Sex-diagnostic-wise distribution of three groups of psychiatric patients

Sl. No.	Disorder	Sex	Psychiatric Patients Population	Government Mental Hospitals	General Hospital Psychiatric Units
1	Organic psychosis	Male (%)	27	**73**	54
2	Substance use disorder	Male (%)	88	**97**	96
3	Schizophrenia	Male (%)	43	**66**	61
4	Mania	Male (%)	**93**	76	65
5	Bipolar affective disorder	Male (%)	43	**74**	62
6	Endogenous depression	Male (%)	32	**57**	46
7	Phobic neurosis	Male (%)	31	**80**	48
8	Anxiety neurosis	Male (%)	28	**61**	60
9	Obsessive compulsive neurosis	Male (%)	30	**63**	62
10	Neurotic depression	Male (%)	20	**62**	49
11	Dissociative disorder	Male (%)	14	**50**	17
12	Somatoform disorder	Male (%)	**62**	60	41
13	Behavioral syndrome	Male (%)	58	**75**	74
14	Personality disorder	Male (%)	90	**95**	60
15	Mental retardation	Male (%)	69	**72**	63
16	Developmental disorder	Male (%)	58	**96**	79
17	Behavioral/ emotional disorder	Male (%)	52	**67**	55
18	Epilepsy	Male (%)	57	**67**	60
	All disorders	Male (%)	46	**69**	41

From the Table 5.12 it can be noted that, in government mental hospitals, the male proportions in 16 out of 18 diagnostic categories are highest. In psychiatric patient population, the male proportions are highest in mania and somatoform disorders only. In general hospital psychiatry units, the male proportion is not highest in any of the diagnostic categories.

5.10   SERVICE INDICATORS OF GENERAL HOSPITAL PSYCHIATRIC UNITS

5.10.1   OUT-PATIENT SERVICE INDICATORS

Out-patient service indicators of general hospital psychiatric units are given in Table 5.13

TABLE 5.13   Out-patient service indicators of general hospital psychiatric units

Out-patient service indicators:		
1. Number of new registrations: Ranged from 59 to 2,514 at individual hospitals		
2. Age (%):	Below 17	23.8
	17–60	70.2
	Above 60	6.0
3. Sex (%)	Male (%):	41.0
	Female (%):	59.0
4. Diagnosis (%):		
	Organic psychoses	4.9
	Substance use disorders	10.7
	Schizophrenia	20.3
	Mania	2.5
	Bipolar affective disorders	4.5
	Endogenous depression	20.8
	Phobic neuroses	1.6
	Anxiety neuroses	7.1
	Obsessive compulsive neuroses:	1.9
	Neurotic depression	4.6
	Dissociative disorders	4.0
	Somatoform disorders	4.5
	Behavioral disorders	2.5
	Personality disorders	0.8
	Mental retardation	5.1
	Developmental disorders	0.7
	Behr./emotional disorders	1.6
	Epilepsy	1.9
5. Rate of follow-up attendance		2.1

5.11 OPTIMIZATION OF SERVICE INDICATORS

The reduction in the heterogeneity of important service indicators among the government mental hospitals will optimize the services rendered by these hospitals in any country. The optimization process has both clinical and economic implications.

5.11.1 REDDY'S METHOD

The optimum service indicators of a government mental hospital may be determined as follows:
1. Minimum number of registrations = Number of in-patients × Rate of registrations of all the hospitals in the country.
    where rate of registration = Number of registrations during the year / Number of in-patients as on first July of the year
2. Total follow-up attendences = Number of new registrations × Rate of follow-up attendance of all the hospitals
3. Number of admissions = Number of registrations × Rate of admissions of all the hospitals
4. Maximum number of chronic patients = Number of in-patients × Propotion of chronic patients of all the hospitals
    Example : In India for 400 in-patients (Bed strength) as on first July of the year, the optimum indicators are calculated as follows.
    Minimum number of new registrations = 400 × 12 = 4,800
    Total follow-up attendance = 4,800 × 4.5 = 21,600
    Number of admissions = 4,800 × 0.38 = 1,440
    Maximum number of chronic patients = 400 × 0.476 = 190

5.12 LONG-STAY PATIENTS IN GOVERNMENT MENTAL HOSPITALS

5.12.1 NEED FOR DATA

Sufficient data on long-stay patients in government mental hospitals are lacking in the world. In the absence of determining the optimum indicators for long-stay patients in the system, the national indicators may serve as optimum values. The hospital indicators are trends that were significantly high and may aid in the planning of rehabilitation facilities and services, and also sets the guidelines for discharge procedures in order to make in-patients service more effective with the existing bed strength. Such a report may also serve as baseline data for evaluating the services rendered by the system at national and individual hospital levels.

5.12.2   LONG-STAY PATIENTS IN GOVERNMENT MENTAL HOSPITAL IN INDIA

The load of long-stay patients (staying for more than 2 years) in 36 government mental hospitals in India as on 1st July 1999 is shown in Table 5.14.

TABLE 5.14   Number of inpatients and number and rate of long-stay patients at government mental hospitals in India

States and Names of Hospitals	In-patients	Long-stay Patients	
		Number	%
All hospitals	15,345	7,307	48
Andhra Pradesh			
Institute of Mental Health, Hyderabad	386	71	16
Institute of Mental Health, Vishakapatnam	300	20	7
Assam			
LGB Regional Institute of Mental Health, Tezpur	353	103	29
Jharkhand			
Ranchi Institute of Neuro-Psychiatry and Allied Sciences, Ranchi	543	191	35
Central Institute of Psychiatry, Ranchi	360	99	28
Delhi			
Institute of Human Behaviour and Allied Sciences, Delhi	140	44	31
Goa			
Institute of Psychiatry and Human Behaviour, Panaji	150	80	53
Gujarat			
Hospital for Mental Health, Ahmedabad	402	247	61
Hospital for Mental Health, Vadodara	181	131	72
Hospital for Mental Health, Jamnagar	55	8	15
Hospital for mental health, Bhuj	25	7	28

TABLE 5.14   *(Continued)*

Jammu and Kashmir			
Psychiatric Diseases Hospital, Srinagar	100	90	90
Karnataka			
National Institute of Mental Health and Neuro-sciences, Bangalore	364	56	15
Karnataka Institute of Mental Health, Dharwad	296	84	28
Kerala			
Government Mental Health Center, Trivandrum	774	305	39
Government Mental Health Center, Thrissur	382	61	16
Government Mental Health Center, Kozhikode	685	381	56
Madhya Pradesh			
Gwalior Mansik Arogyashala, Gwalior	192	38	20
Mental Hospital, Indore	157	107	68
Maharashtra			
Regional Mental Hospital, Thane	1,744	835	48
Regional Mental Hospital, Pune	2,540	1,848	73
Regional Mental Hospital, Nagpur	786	470	60
Regional Mental Hospital, Ratnagiri	183	78	43
Nagaland			
Mental Hospital, Kohima	21	9	43
Punjab			
Punjab Mental Hospital, Amritsar	415	314	76
Rajasthan			
Psychiatric Centre, Jaipur	312	56	18
Tamil Nadu			
Institute of Mental Hospital, Chennai	1,657	692	42
Uttar Pradesh			
Agra Mansik Arogyashala, Agra	459	142	31
Mental Hospital, Varanasi	258	133	52
Mental Hospital, Bareilly	292	192	66
West Bengal			
Calcutta Mansik Hospital, Calcutta	251	175	70
Lumbini Park Mental Hospital, Calcutta	129	60	47
Institute of Psychiatry, Calcutta	36	2	6
The Mental Hospital, Mankundu	106	10	9
Berhampore Mental Hospital, Berhampore	214	79	37
Institute of Mental Care, Purulia	97	89	92

KEYWORDS

- **Demography**
- **Mental morbidity**
- **Prevalence rate**
- **Service indicators**

CHAPTER 6

PROBABILITY AND PROBABILITY DISTRIBUTIONS

CONTENTS

6.1   PROBABILITY SCALE

The probability is the degree of confidence in the occurrence of events. It is the measurement of events. This measure being a number lies between 0 and 1. Thus,

$P = 0$ implies that the occurrence of the event is impossible

$P = 1$ implies that the occurrence of the event is certain

$P = 0.5$ implies that 50 percent chance for the event to occur.

6.1.1   DEFINITIONS OF PROBABILITY

6.1.1.1   THEORETICAL PROBABILITY

The theoretical (mathematical, classical, a priori) probability of occurrence of an event is given by

$$P(E) = \frac{\text{Number of favorable cases}}{\text{Number of possible cases}}$$

Example: In tossing a coin, there are two possible cases: occurrence of head and occurrence of tail. Let us suppose that the occurrence of head is favorable to us. Then,

$P(\text{head}) = \dfrac{1}{2} = 0.5$ or 50%

6.1.1.2   PRACTICAL PROBABILITY

The practical (statistical, empirical, frequency, a posteriori) probability is given by

$$P(E) = \frac{\text{Number of times the event has occured}}{\text{Total number of trials}}$$

Example: There are 24 males in a total of 40 psychiatric registrations at NIMHANS hospital (data in Appendix I).Then, the probability that a registered patient happens to be a male is obtained by

$$P(\text{male}) = \frac{24}{40} = 0.60 \text{ or } 60\%$$

6.1.2 LAWS OF PROBABILITY

6.1.2.1   ADDITION LAW

If two events A and B are mutually exclusive, then the probability of the event "A or B" is given by the sum of their probabilities. That is,

$$P(A \text{ or } B) = P(A) + P(B)$$

Example: It is found that 20 percent of psychiatric registrations at NIMHANS hospital are children (aged below 17 years) and that 10 percent of the registrations are aged (above 60 years). Then, the probability that a registered patient happens to be either a child or aged is obtained as follows:

$$P(\text{child or aged}) = P(\text{child}) + P(\text{aged})$$
$$= 0.20 + 0.10 = 0.30 \text{ or } 30\%$$

6.1.2.2   MULTIPLICATION LAW

If two events A and B are mutually independent, then the probability of the event "A and B" is given by the product of their probabilities. That is,

$$P(A \text{ and } B) = P(A) \times P(B)$$

Example: Let us consider two consecutively registered psychiatric patients at NIMHANS registration counter. The probability that both of them happen to be males is obtained (data in Appendix I) as

$P(\text{male and male}) = 0.6 \times 0.6 = 0.36 \text{ or } 36\%$

6.1.2.3   COMPLEMENTARY PROBABILITY

If A is an event, not-A is called its complementary event, and it is denoted by A'. Its probability called complementary probability is given by

$$P(A') = 1 - P(A)$$

Example: The probability that a registered patient at NIMHANS hospital happens to be a female is given by

$$P(\text{female}) = 1 - P(\text{male}) = 1 - 0.60 = 0.40 \text{ or } 40\%$$

6.1.2.4 CONDITIONAL PROBABILITY

If the event A can occur only when it is known that B has occurred, then the occurrence of A is conditional to the occurrence of B. The conditional probability of A given that B has already occurred is obtained by

$$P(A/B) = \frac{P(AB)}{P(B)}$$

where P(AB) is the probability of occurrence of the event.

Example: It is found that 0.4 percent ($P = 0.004$) of psychiatric registrations at NIMHANS hospital have puerperal psychosis and that 40 percent ($P = 0.4$) of the registered patients are females. Then, the probability that a registered patient has puerperal psychosis given that the patient is a female is obtained by

$$P(\text{puerperal psychosis/female}) = \frac{0.004}{0.4} = 0.01 \text{ or } 1\%$$

6.1.3 BAYES' THEOREM

In general, the psychiatrist knows the probability of occurrence of a particular symptom in a patient with a particular psychiatric disorder (d). However, it is important to know the probability of occurrence of a particular psychiatric disorder in a patient who has a particular symptom. A theorem attributed to Thomas Bayes may be used to provide the latter probability from the former probability.

6.1.3.1 FORMULA

It is given by

$$P(d/s) = \frac{P(d) \times P\left(\frac{s}{d}\right)}{P(d) \times P\left(\frac{s}{d}\right) + P(d') \times P\left(\frac{s}{d'}\right)}$$

Example: It is found that 28 percent of children registered at CGC of NIMHAMS hospital have conduct disorder. Consider the two events that a child has conduct disorder and that the child has non-conduct disorder. It is further known that 63 percent of the conduct disorder children and 17 percent of the non-conduct disorders children of the clinic have truancy at home/school. Then, the probability that a registered child has conduct disorder knowing that the child has truancy is given by

$$P(\text{conduct disorder/truancy}) = \frac{0.28 \times 0.63}{(0.28 \times 0.63) + (0.72 \times 0.17)} = \frac{0.1764}{0.2988} = 0.59 \text{ or } 59\%$$

6.1.4 EVALUATION OF SCREENING TESTS

6.1.4.1 SENSITIVITY AND SPECIFICITY

The sensitivity of screening test is the ability of the test to detect positive cases. That is, the sensitivity is the probability that the screening test indicates a diseased case as positive. The specificity of screening test is the ability of the test to detect negative cases. That is, the specificity is the probability that the screening test indicates a non-diseased case as negative.

A 2 × 2 screening test table

Screening test	Disease	Non-disease
Positive	True positive (TP)	False positive (FP)
Negative	False negative (FN)	True negative (TN)
Total	(D)	(ND)

Now,

$$\text{Sensitivity of the test} = \frac{\text{TP}}{\text{(D)}}$$

$$\text{Specificity of the test} = \frac{\text{TN}}{\text{(ND)}}$$

6.1.5 DEALING WITH CONFIDENTIAL INFORMATION

The estimation of parameters in case of sensitive (confidential) information is unduly affected by the presence of high rate of false answers. This requires an indirect way of obtaining data for a reliable estimate of parameters.

Example: Let us suppose that we are interested in estimating the probability of premarital sex among married women at a psychiatric clinic.

A: Select a random number from 1 to 10. If it is greater than 3, then answer to question B. Otherwise, answer to question C.

B: Have you had premarital sex?

C: Toss a coin. Is it head?

Let us suppose that there are 25 'yes' responses in a total of 100 trials. Then,

$$\frac{25}{100} = P(B \text{ and Yes}) + P(C \text{ and Yes})$$

$$= 0.7 \, P + 0.3 \times 0.5$$
$$= 0.7 \, P + 0.15$$
$$0.7 \, P = 0.25 - 0.15 = 0.10$$

$$P = \frac{0.10}{0.70} = 0.143 \text{ or } 14.3\%$$

6.2 PROBABILITY DISTRIBUTIONS

The statistical inferences are drawn by considering sampling distributions and calculating probabilities. The sampling distributions differ according to the type of the characteristic studied, the nature of the population, and the size of the sample. For each type of situation, a sampling distribution may be formed by using a mathematical model called theoretical distribution. In several situations, the observed sampling distributions are very close approximation of the theoretical distributions. The mathematical models have been developed, and the required probabilities were calculated and made available by means of tables for certain types of distributions.

6.2.1 BINOMIAL DISTRIBUTION

Sometimes, the psychiatrists are interested to know the proportion of individuals in population possess a particular character such as the number of psychiatric patients in a family of two members. An estimate of this proportion is calculated based on a suitably drawn sample and the corresponding sampling distribution. In this type of problem, the sampling distribution is given by a theoretical frequency distribution known as bi-nominal distribution. It is given by

$$P(x) = n \, C_x \, p^x \, q^{n-x}$$

where

n is the sample size

x is the number of individuals who possess the event

p is the probability of occurrence of the event

$$q = 1-p$$

The nC_x denotes the number of combinations in a set of n objects taken x at a time. It is given by

$$nC_x = \frac{n!}{(n-x)!x!}$$

where $n! = 1 \times 2 \times \ldots \times n$

Example: It is found that the probability of a male registered patient at NIMHANS hospital is 0.60. A sample of two patients at the registration counter can be any one of three types: having no male patient, one male patient, or two male patients. The probabilities of these can be obtained by using binominal probability distribution formula as shown below:

$P(O) = 2C_0(0.6)^0(0.4)^2 = 1 \times 1 \times 0.16 = 0.16$

$P(1) = 2C_1(0.6)^1(0.4)^1 = 2 \times 0.6 \times 0.4 = 0.48$

$P(2) = 2C_2(0.6)^2(0.4)^0 = 1 \times 0.36 \times 1 = 0.36$

6.2.2 POISSON DISTRIBUTION

There are situations in which the number of times an event occurs can be counted, but the number of times the event did not occur cannot be count-ed. For example, the number of follow-ups made by a psychiatric patient may be counted, but this is not possible when the number of follow-ups is not made. The probabilities of observing no follow-up, one follow-up, two follow-ups, and so on, in a given sample of such patients can theoretically be found out by the using Poisson distribution. It is given by

$$P(x) = \frac{e^{-m}m^x}{x!}$$

where x is the number of times the event occurred

m is the mean of the distribution

6.2.3 NORMAL DISTRIBUTION

In many situations, the characteristics (variables) to be studied are con-tinuously measured such as intelligence quotient or IQ, body temperature, etc. For such variables, many populations, and also their sampling dis-tributions, are very close to a pattern of frequency distribution known as normal distribution. The normal distribution is denoted by $N(\overline{x}, \sigma^2)$. The mathematical function that generates the probabilities is given in the lit-erature. An ideal normal distribution curve is as shown in Figure 6.1.

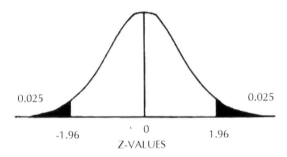

FIGURE 6.1   Probability of standard normal values of an observation falling ± 1.96.

An ideal normal distribution curve is symmetrical. About 95 percent of the observations lie within 1.96 standard error from either side of the mean, and about 99 percent of the observations lie within 2.58 standard error from either side of the mean.

6.2.3.1 CENTRAL LIMIT THEOREM

The central limit theorem states that the statistics of all distributions follow normal distribution provided the sample size is sufficiently large. Many variables encountered in psychiatric research have approximate normal distribution. If the distribution is approximately normal, no essential de-tails are lost by considering it to be normal. The mean and standard devia-tion describe the normal distribution completely.

6.2.4 STANDARD ERRORS OF STATISTICS

The sampling distribution of the statistics calculated from samples taken from the parent normal distribution is known. If a series of samples of fixed size is taken, say n from the population, then the sample means cluster closely around the population mean. The standard deviation of statistic is known as standard error (SE). The SE decreases with an increase in the sample size.

6.2.4.1 STANDARD ERROR OF SAMPLE MEAN

It is given by

$$SE(\overline{x}) = \frac{s}{\sqrt{n}}$$

Example: The SE of the mean of the marks scored by five students (7, 4, 1, 3, and 5) is calculated as follows:

$$SE(^{-}) = \frac{\sqrt{5}}{\sqrt{5}} = 1$$

6.2.4.2 STANDARD ERROR OF SAMPLE PROPORTION

It is given by

$$SE(p) = \sqrt{\frac{pq}{n}}$$

Example: The standard error of the proportion of males ($p = 0.6$) in total of 40 registered patients (data in Appendix I) is calculated as follows:

$$SE(p) = \sqrt{\frac{(0.6 \times 0.4)}{40}} = \sqrt{0.006} = 0.077$$

6.2.5 DERIVED DISTRIBUTIONS

Several probability distributions have been developed from the parent normal distribution.

6.2.5.1   STANDARDIZED VALUES

If x is a normal variate, then the standard normal variate or relative normal variate (denoted by Z) is defined as follows:

$$Z = \frac{(x - \bar{x})}{S}$$

The mean of the standard normal values in a dataset is 0 and the standard deviation is 1. The standardized distribution is denoted by $N(0,1)$. The proportions under the entire standard normal curve that lies between 0 and a positive value of Z are presented in probability distribution tables (Appendix III). The proportion of areas between zero and a negative value of Z are obtained by symmetry.

Example: The standardized marks scored by the five students (7, 4, 1, 3, and 5) are computed as shown in the following table.

Student	x	$(x - \bar{x})$	Z
1	7	3	1.5
2	4	0	0
3	1	−3	−1.5
4	3	−1	−0.5
5	5	1	0.5
Mean	4	0	0

6.2.5.2 χ^2(chi-square) DISTRIBUTION

It is the sum of squares of n independent standard normal variates. Given a fixed total in a contingency table, the number of independent cells that can be varied freely will be the number of degrees of freedom (df).

6.2.5.3   t-DISTRIBUTION

It is the ratio of a standard normal variate to the square root of the χ^2 divided by their respective degrees of freedom. The size of the sample minus the number of parameters estimated is called the degree of freedom. Thus, it is the number of independent members in the sample.

6.2.5.4 F-DISTRIBUTION

It is the ratio of two independent χ^2 divided by their respective degree of freedom. Thus, the distribution has two independent degrees of freedom: one attached to the numerator and the other attached to the denominator.

KEYWORDS

- **Bayes, theorem**
- **Sampling distribution**
- **Screening tests**
- **Standard error**

CHAPTER 7

SAMPLING THEORY AND METHODS

CONTENTS

7.1 THEORY OF SAMPLING

7.1.1 RANDOM SAMPLING

Generally, we are interested in representative sample to draw valid conclusions on the population. Such representation can be assured only through random sampling. In random sampling, every unit in the population has a known probability of being included in the sample. The sample is drawn by some method of random selection consistent with these probabilities, and we take into account of these probabilities while estimating the parameters. In random sampling, it is possible to study the bias and error attached to the estimates from different sampling techniques. In this way, much has been reported about the scope, advantages, and limitations of each sampling technique so as to choose the one that suits our sampling job reasonably well.

7.1.2 NON-RANDOM SAMPLING

Non-random sampling techniques such as convenient sampling, judgment sampling, and quota sampling do not assure fair representation of the population, and hence they must be based on certain specific purposes.

7.1.3 SAMPLING ERROR

In random sampling, there are sampling errors due to sample enumeration instead of complete enumeration (census enumeration). That is, the sampling errors are the errors of estimates that arise solely from the sampling variation that is present whenever a sample of n units (sample) is measured instead of the complete population of N units (population size). The sampling errors may be reduced by selecting suitable random sampling method, fixing optimum sampling size, and eliminating sample bias.

7.1.4 SAMPLING BIAS

The sampling bias is nothing but consistent errors that arise due to the sample selection. It may be due to wrong demarcation of the sample from the population, measuring the succeeding or proceeding units in the list, and the errors of estimation. Sampling bias means that the data collected may not represent one group.

7.2 RANDOM SAMPLING METHODS

Several random sampling methods are available depending on the objectives of the study, the size and nature of the population, required precision of the estimates, sanctioned budget, and the availability of the sampling frame. A sampling frame is a list of all the units in the population listed alphabetically or chronologically and numbered serially.

7.2.1 SIMPLE RANDOM SAMPLING

This is the simplest of all the random sampling methods. In simple random sampling, each unit in the population has the same chance of being included in the sample at the first draw or at each subsequent draws. The sampling may be carried out by using random number tables or by lottery system of drawing. This sampling method is appropriate when the population size is small, sampling units are homogeneous, and the frame is readily available as in the case of clinical trials.

7.2.1.1 ESTIMATE

An unbiased estimate of the population mean is given by

$$\overline{y}_{SRS} = \frac{1}{n}\sum y_i$$

where n is the sample size
y_i is the observed value of the ith observation
Random numbers: A list of random numbers is given below:

6	8	2	3	6	0	3	9	2	8	9	1	2	3	9	3	8	9	6	4	1	5	3
1	3	2	3	6	5	4	5	2	3	7	9	1	0	5	3	2	7	2	1	5	1	1
0	5	2	4	3	8	6	7	1	3	2	9	2	3	2	5	2	7	9	3	6	5	7
9	0	5	1	0	9	8	9	6	8	0	3	7	9	2	0	5	1	5	1	6	9	4

7.2.2 SYSTEMATIC SAMPLING

The systematic sampling is operationally more convenient than simple random sampling and at the same time ensures equal probability to each unit in the population of being inclusion in the sample. Let $k = N/n$, where N is the population size. k is taken as the nearest integer to N/n. Select a random number from 1 to k, say r. Then, the rth, $(r + k)$th, $\{r + (n - 1)k\}$ th units are selected for the sample. The k is known as sampling interval that is the reciprocal of the sampling fraction, and the random number r is known as random start. This sampling procedure is not valid when there is a periodicity of a particular event under study in the population.

7.2.2.1   ESTIMATE

An unbiased estimate of the population mean is given by

$$\bar{y}_{SYS} = \frac{1}{n}\sum y_i$$

7.2.3   STRATIFIED SAMPLING

Stratified sampling may be preferred to increase the precision of the estimate by reducing the heterogeneity of the population. In stratified sampling, the population of N units is divided into L sub-populations of $N_1, N_2,$ ... N_L units that are internally homogeneous. These sub-populations are not overlapping, and together they comprise the whole population. The classes into which the population is divided are called strata. Then, each stratum will be sub-sampled, and a definite number of units are taken from each stratum as in the case of simple random sampling. The partial samples of the different strata may be used to estimate the means of several strata from which the overall estimate of the population mean may be obtained.

7.2.3.1 ESTIMATE

An unbiased estimate of the population mean is given by

$$\bar{y}_{ST} = \sum W_h \bar{y}_h$$

where

$$W_h = \frac{N_h}{N}$$

$$\bar{y}_h = \frac{1}{n_h} \sum y_{hi}$$

The n_h is the sample size for the h^{st} stratum. Various allocation models are available in fixing the sample size for each stratum, such as proportional allocation model, etc.

7.2.4 CLUSTER SAMPLING

It consists in forming suitable clusters of units and surveying all the units in a sample of selected cluster. By cluster sampling, it means sampling of clusters of units formed by grouping neighboring units or units that can be surveyed together. For example, while surveying a random sample of schools in a particular city to determine scholastic backwardness of first-grade students in the city. The units within the clusters should be as heterogeneous as possible. Let us deal with the case in which all the clusters are of equal size. Suppose a finite population of NM units is divided into L clusters of size M in each cluster. A sample of n clusters is selected from N clusters using simple random sampling without replacement.

7.2.4.1 ESTIMATE

An unbiased estimate of the population mean is given by

$$\overline{y}_c = \frac{1}{n} \sum \overline{y}_i$$

where $\overline{y}_i = \dfrac{1}{M} \sum y_{ij}$

7.2.5 AREA SAMPLING

Area sampling is close to cluster sampling, which is more suitable for mental morbidity surveys. Under area sampling, we first divide the total area into a number of smaller non-overlapping areas called geographical clusters, then a number of these clusters are randomly selected, and all the units in the selected clusters are included in that sample.

7.3 RANDOM SAMPLING METHODS IN MENTAL MORBIDITY SURVEYS

The mental morbidity surveys in India adapted the following schemes of selection of study areas/sampling procedures depending on the nature of the population (general or specific).

7.3.1 RURAL SECTORS

For rural sectors, the schemes of selection of study areas/sampling procedures are the following:
- A part of the village
- A single village
- Villages in two wards of a *panchayat*
- All families of three communities in clusters of small villages
- A random sample of families from four villages
- A random sample of 50 percent of families in 11 villages of a block

7.3.2 URBAN SECTORS

For urban sectors, the schemes of selection of study areas/sampling procedures are the following:
- A random sample of families from urban health center
- A slum locality in a metropolitan city

7.3.3 SEMI-URBAN SECTORS

For semi-urban sectors, the schemes of selection of study areas/sampling procedures are as follows:
- A sample of families of officers staying in their official quarters in a particular area of a city
- A sample of families of a tribal community migrated to an urban area

KEYWORDS

- **Sampling bias**
- **Sampling methods**

CHAPTER 8

BASIC ELEMENTS OF STATISTICAL INFERENCE

CONTENTS

8.1 BASIC ELEMENTS OF ESTIMATION OF PARAMETERS

An estimator is a function of sample values, whereas an estimate is a numerical value of the estimator. An estimate may be a single statistic (such as the sample mean for a population mean, sample proportion for a population proportion, etc.), which is called point estimates or a range with attached probabilities called interval estimates or confidence intervals.

8.1.1 POINT ESTIMATES

The criteria for selecting a suitable estimator for a purpose have to be described.

8.1.1.1 CONSISTENCY, EFFICIENCY, AND SUFFICIENCY

A consistent estimator gets increasingly better as the sample size increases. For example, the sample mean and the median are consistent estimators of their respective parameters. An efficient estimator is one for which the standard error is small. In many situations, the sample mean tends to be more efficient estimator than the sample median, although there are situations for which the reverse is also true. A sufficient estimator makes use of all the potentially useful information in the data. Thus, the sample mean is a sufficient estimator, but not the sample median.

8.1.1.2 RESISTANCE

A resistant estimator is not affected too much by the presence of extreme values or the outliers in the data. For example, the sample median is a resistant estimator, but not the sample mean. We should pay particular attention to resistant estimator in psychiatric research since gross errors or outliers occur frequently in practice.

8.1.1.3 UNBIASED

The expected value of an unbiased estimator is equal to the parameter value. The expected value of a random variate is given by

$$E(x) = \sum_i x_i p_i$$

where x_i are the sampling points

p_1 are the attached probabilities of the sampling points

Example: The expected value when a die is rolled is calculated as follows:

$$E\,(\text{die}) = (1 + 2 + 3 + 4 + 5 + 6)/6 = 21/6 = 3.5$$

The sample mean $(\bar{x})$ is an unbiased estimator of the population mean (μ). An unbiased estimator of standard deviation of sample is given by

$$s = \sqrt{\frac{1}{(n-1)}\sum(x - \bar{x})^2}$$

Example: The unbiased standard deviation for five values (7, 4, 1, 3, and 5) is calculated as follows:

$$s = \sqrt{\frac{20}{4}} = \sqrt{5} = 2.236$$

Whereas the biased estimator for the same data has been calculated as follows:

$$S = \sqrt{\frac{20}{5}} = \sqrt{4} = 2.0$$

8.1.2 INTERVAL ESTIMATES

The estimation of confidence interval (CI) is important since the investigator can never estimate the exact values of the parameter with certainty. The procedure is to obtain a point estimate and then set up certain limits on both sides of the estimate on the basis of the sampling distribution of the statistics used. Thus, the CI helps in locating the parameter values. The sample mean $(\bar{x})$ is the best point estimator for population mean (μ).

8.1.2.1   CI FOR MEAN

When the sample size is large (more than 30), the 95 percent CI for estimating population mean is given by

$$\bar{x} \pm 1.96 \times SE(\bar{x})$$

That is,

$$\bar{x} \pm 1.96 \times \frac{s}{\sqrt{n}}$$

when the sample size is small, the 95 percent CI is given by

$$\bar{x} \pm t \times SE(\bar{x})$$

Example: The 95 percent CI for the marks scored by the five students
(7, 4, 1, 3, and 5) is worked out as follows:

$$4 \pm 2.78 \times 1, \text{ that is } 1.22 \text{ to } 6.78$$

where the table value of $t(4) = 2.78$, at 5 percent level of significance.

8.1.2.2   CI FOR PROPORTION

The sample proportion (p) is the best estimator of the population proportion (P). When the sample size is large, the 95 percent CI for estimating
population proportion is given by

$$p \pm 1.96 \times SE(p)$$

That is,

$$p \pm 1.96 \times \sqrt{\frac{pq}{n}}$$

Example: The 95 percent CI for population proportion of male patients at NIMHANS registration (data in Appendix I) is worked out as follows:

$$0.6 \pm 1.96 \times 0.077$$

$$0.6 \pm 0.15, \text{ that is, } 0.45 \text{ to } 0.75$$

8.1.3   DETERMINATION OF SAMPLE SIZE

The sampling variance decreases with increase in the sample size; but at the same time, the cost of the survey also increases. Hence, in practice, the optimum sample size that decreases the sampling error for a reasonable cost has to be determined. The general relationship between the sample size and sampling error is as shown in the Figure 8.1.

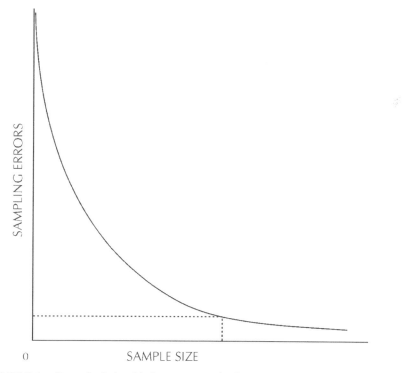

FIGURE 8.1   General relationship between sample size and sampling error.

8.1.3.1   SAMPLE SIZE FOR ESTIMATING POPULATION MEAN

In case of estimating a population mean, the optimum sample size is determined by

$$n = \frac{Z_{\alpha}^{2}S^{2}}{d^{2}}$$

where Z_{α} is fixed at 1.96 for 5 percent level of significance

S^{2} is the estimate of the variance from a pilot study

d is the allowable error (difference from the true value) due to sampling defects that can be tolerated. It is usually fixed at 10 percent of the available estimate from the pilot study.

Example: Let us support that the marks scored by a randomly selected five students in a test are 7, 4, 1, 3, and 5. Then, the sample size required for estimating the population mean is determined as follows:

$$n = \frac{1.96^{2} \times 5}{(4 \times 0.1)^{2}} = \frac{3.841 \times 5}{0.16} = \frac{19.205}{0.16} = 120$$

8.1.3.2   SAMPLE SIZE FOR ESTIMATING POPULATION PROPORTION

In case of estimating a population proportion, the optimum sample size is determined as follows:

$$n = \frac{Z_{\alpha}^{2}pq}{d^{2}}$$

where p is the available estimate of the proportion

Example: The sample size required for estimating the proportion of males at NIMHANS psychiatric registrations (data in Appendix I) is given by

$$n = \frac{1.96^{2} \times 0.6 \times 0.4}{(0.6 \times 0.1)^{2}} = \frac{3.841 \times 0.24}{0.0036} = 256, \text{ or approximately } 260$$

8.2   BASIC ELEMENTS OF TESTS OF SIGNIFICANCE

A researcher involves in making hypothesis about population and checking the plausibility of the hypothesis using sample data. The sampling variation is considered in this process.

8.2.1   NULL AND ALTERNATE HYPOTHESES

The assumption of no difference between populations or no association between factors in the population is known as null hypothesis and is denoted by H_0. When the null hypothesis is rejected, the other hypothesis $(H_1, H_2, \ldots)$ known as alternate hypothesis has to be accepted.

8.2.2   STATISTICAL SIGNIFICANCE

This means that the difference between the observed score from the sample and the score based on the hypothesis is not due to chance, defying the null hypothesis. On the contrary, the statistic is not significant means that the difference is obtained by chance.

8.2.3   TYPE I AND TYPE II ERRORS

Type I error rejects the null hypothesis when it is actually true, whereas type II error accepts the null hypothesis when it is actually false. A table showing the type I error and type II error is as given below:

	H_0 is true	H_0 is false
Action taken: Reject	Type I error	-
Accept	-	Type II error

8.2.4   LEVEL OF SIGNIFICANCE AND POWER OF THE TEST

The probability of committing type I error is known as the level of significance. That is,

Level of significance = P (type I error)

It is usually fixed at 5 percent ($P = 0.05$). The power of the test is the probability of rejecting the false null hypothesis. Thus, the power of the test is given by

Power of the test = $1 - P$ (type II error)

Thus, the power of the test is the probability of rejecting the false null hypothesis. Experience suggests that in practice the power of the test is set at a maximum of 80 percent.

8.2.5   TWO-TAILED AND ONE-TAILED TESTS

In test of significances when we want to determine, for example, whether the mean age of onset of schizophrenia is different from that of bipolar affective disorders, and does not specify higher or lower, the P-values include both sides of extreme result, and the test is called two-tailed test. In case of 5 percent level of significance, the P-values will be 2.5 percent at each end. Therefore, the result is compared with the table value at the probability level of 0.05. The difference is being tested for significance, but the direction is not specified. When we want to test, for example, whether the mean of schizophrenia is less than that of bipolar affective disorders, the results will be at one end of the tail of the distribution, and the test is called one-tailed test.

8.2.6   TEST STATISTIC

In statistical hypothesis testing, a hypothesis test is typically specified in terms of test statistic that is a function of the sample. It is considered as a numerical summary of a set of data that reduces the data to one or a small number of values that can be used to perform a hypothesis test.

8.2.7   STEPS INVOLVED IN STATISTICAL TESTS OF SIGNIFICANCE

There are various types of problems for which the test of significance are used for drawing conclusions. Different types of problem require different tests, but the basis and the steps involved in the procedures are one and the same.

1. Stating the null hypothesis and the alternate hypotheses.
2. Formulate statistic and calculate its standard error. That is, calculate the critical ratio and determine its distribution.
3. Compare the observed value of the critical ratio with that of the table value at 5 and 1 percent level of significance. If the calculated value is less than the table value, then accept the null hypothesis. If the calculated value is more than the table value, then reject the null hypothesis and accept the alternate hypothesis.

KEYWORDS

- **Confidence interval**
- **Point estimates**
- **Resistant estimator**
- **Statistical tests**

CHAPTER 9

PARAMETRIC TESTS OF SIGNIFICANCE

CONTENTS

9.1   ONE-SAMPLE PARAMETRIC TESTS

9.1.1   ONE-SAMPLE *t*-TEST

9.1.1.1   HYPOTHESES

The one-sample *t*-test is applied to test for a population mean. It compares the mean score of a sample ($\overline{x}$) to an unknown value (μ_0). The unknown value is a population mean. That is,

H_0: $\mu = \mu_0$ against H_1: $\mu \neq \mu_0$

9.1.1.2   TEST STATISTIC

It is given by

$$t = \frac{|\,\overline{x} - \mu_0\,|}{s / \sqrt{n}} \quad \text{with df} = (n{-}1)$$

9.1.1.3   SAMPLE SIZE

The optimum sample size is determined by

$$n = \frac{\left(Z_\alpha + Z_\beta\right)^2 S^2}{\left(\overline{x} - \mu_0\right)^2}$$

where $Z_\alpha = 1.96$ and $Z_\beta = 0.84$

Example: Let us assume that the improvement scores of a group of five patients in a test are given by 7, 4, 1, 3, and 5. We wish to test whether the mean score of the whole class can be considered as 5.0. That is,

$$t = \frac{|4 - 5|}{\sqrt{5} / \sqrt{5}} = 1.0, \text{ with df} = (5{-}1) = 4$$

where $s = \sqrt{5}$

Since the calculated value is less than the table value of $t(4) = 2.78$ at 5 percent level of significance, we accept the null hypothesis ($P > 0.05$, not significant).

For this data,

$$n = \frac{(1.96 + 0.84\ )^2 \times 5}{(4 - 5)^2} = 2.8^2 \times 5 = 39.2$$

Thus, the sample size required for this purpose is about 40.

9.1.2   ONE-SAMPLE PROPORTION TEST

9.1.2.1   HYPOTHESES

The one-sample proportion test is applied to test for a population proportion. It compares the sample proportion (p) to a known proportion (P_0). That is,

$$H_0{:}P = P_0 \text{ against } H_1{:} P \neq P_0 \text{ (large sample)}$$

9.1.2.2   TEST STATISTIC

This is given by the following relation:

$$Z = \frac{|\,p - P_0\,|}{\sqrt{P_0 Q_0 / n}}$$

where n is the sample size

p is the observed sample proportion

P_0 is the proportion under the null hypothesis

$$Q_0 = 1 - P_0$$

9.1.2.3   SAMPLE SIZE

In testing for a population proportion, the optimum sample size is determined by the following relation:

$$n = \left(\frac{Z_\alpha \sqrt{P_0 Q_0} + Z_\beta \sqrt{pq}}{(p - P_0)} \right)^2$$

Example: A sample of 40 registrations at NIMHANS showed that there are 24 males ($p = 0.6$). We wish to test whether the population proportion can be considered as 0.5. That is,

$H_0: P_0 = 0.50$, against $H_1: P_0 \neq 0.50$.

For this data,

$$Z = \frac{|0.60 - 0.50|}{\sqrt{(0.5 \times 0.5)/40}} = \frac{0.10\sqrt{40}}{0.5} = 1.26$$

Since the calculated value is less that 1.96, we accept the null hypothesis ($P > 0.05$, not significant).

For this data,

$$n = \left(\frac{1.96\sqrt{0.5 \times 0.5} + 0.84\sqrt{0.6 \times 0.4}}{(0.6 - 0.5)} \right) = \left(\frac{0.98 + 0.41}{(0.1)} \right)^2 = 193$$

Thus, the sample size required for the purpose is about 200.

9.2 TWO-INDEPENDENT SAMPLE PARAMETRIC TESTS

9.2.1 INDEPENDENT SAMPLE t-TEST

9.2.1.1 HYPOTHESES

The independent sample *t*-test is applied to test for equality of two population means. It compares the mean scores of two independent samples. That is,

$H_0: \mu_1 = \mu_2$ against $H_1: \mu_1 \neq \mu_2$

9.2.1.2 TEST STATISTIC

It is given by

$$t = \frac{|\bar{x} - \bar{y}|}{\sqrt{\frac{(\sum x^2 - n_1\bar{x}^2) + (\sum y^2 - n_2\bar{y}^2)}{(n_1 + n_2 - 2)} \times \left(\frac{1}{n_1} + \frac{1}{n_2}\right)}}$$

with df = $(n_1 + n_2 - 2)$

9.2.1.3   SAMPLE SIZE

The optimum sample size is given by

$$n = \frac{\left(Z_\alpha + Z_\beta\right)^2 S^2}{\left(\bar{x} - \bar{y}\right)^2}$$

Example: Let us assume that the improvement scores of five randomly selected patients under treatment to be 7, 4, 1, 3, and 5, respectively. The improvement scores of another five randomly selected patients under controls are given by 4, 2, 3, 0, and 6. Here, we wish to test,

H_0: $\mu_1 = \mu_2$ against H_1: $\mu_1 \neq \mu_2$.

$$t = \frac{|4-3|}{\sqrt{\frac{(100 - 5 \times 4^2) + (65 - 5 \times 3^2)}{(5+5-2)} \times (\frac{1}{5} + \frac{1}{5})}} = \frac{1}{\sqrt{\frac{20+20}{8} \times \frac{2}{5}}} = \frac{1}{\sqrt{2}} = 0.707$$

Since, the calculated value is less than the table value of $t(8) = 2.31$ at 5 percent level of significance, we accept the null hypothesis.

For this purpose,

$$n = \frac{(1.96 + 0.84)^2 \times 5}{(4-3)^2} = 2.8^2 \times 5 = 39.2$$

The sample size required for this test of significance is about 40.

9.2.2   TWO-INDEPENDENT SAMPLE PROPORTION TEST

9.2.2.1   HYPOTHESES

The two-sample proportion test is applied to test for equality of two populations proportions. It compares the proportions of two independent samples. That is,

$H_0: P_1 = P_2$ against $H_1: P_1 \neq P_2$ (large samples)

9.2.2.2   TEST STATISTIC

It is given by

$$Z = \frac{|p_1 - p_2|}{\sqrt{pq\left(\dfrac{1}{n_1} + \dfrac{1}{n_2}\right)}}$$

where $p = \dfrac{n_1 p_1 + n_2 p_2}{(n_1 + n_2)}$ $q = 1 - p$

9.2.2.3   SAMPLE SIZE

In testing for equality of two population proportions, the optimum sample size is determined by

$$n = \left[\frac{Z_\alpha \sqrt{2 p_1 q_1} + Z_\beta \sqrt{p_1 q_1 + p_2 q_2}}{(p_1 - p_2)}\right]^2$$

Example: There are 24 males in a random sample of 40 registrations at NIMHANS hospital ($p_1 = 0.60$). There are 28 males in a sample of 40 registrations at KIMHANS (Karnataka Institute of Mental Health and Neroscience, Dharward). That is, ($p_2 = 0.70$). Therefore, we wish to test whether the proportions of male registrations at these two institutes are equal. That is,

$$Z = \frac{|0.6 - 0.7|}{\sqrt{(0.65 \times 0.35)(\dfrac{1}{40} + \dfrac{1}{40})}}$$

$$= \frac{0.1}{\sqrt{0.2275 \times 0.05}} = 0.94$$

Since, the calculated value is less than 1.96, we accept the null hypothesis ($P > 0.05$ not significant).

For this data,

$$n = \left[\frac{1.96\sqrt{2 \times 0.6 \times 0.4} + 0.84\sqrt{(0.6 \times 0.4) + (0.7 \times 0.3)}}{(0.6 - 0.7)} \right]^2 = \left[\frac{1.3579 + 0.5635}{-0.1} \right]^2 = 369$$

Thus, the sample size required for the purpose is about 370.

9.3   TWO-RELATED SAMPLE PARAMETRIC TESTS

9.3.1   *PAIRED SAMPLE t-TEST*

9.3.1.1   *HYPOTHESES*

It computes the difference between the two sets of scores (scores before giving the treatment and scores after giving the treatment) for each case and tests to see if the average difference is significantly different from 0. That is,

$H_0: \overline{d} = 0$ against $H_1: \overline{d} \neq 0$

where $d = (x - y)$

x are the scores of the persons before giving the treatment

y are the scores of the persons after giving the treatment

9.3.1.2   *TEST STATISTIC*

It is given by

$$t = \frac{\overline{d}}{s_d / \sqrt{n}} \qquad \text{with df} = (n - 1)$$

where s_d is the unbiased estimate of standard deviation of the differences.

Example: Let us support that the improvement scores of a group of five patients before giving the treatment and the scores after giving the treatment are given below:

Patient	1	2	3	4	5
Before	4	2	3	0	6
After	7	4	1	3	5

The computation table is as given below:

Patient	Before	After	Difference (d)	$(d-\bar{d})$	$(d-\bar{d})^2$
1	4	7	3	2	4
2	2	4	2	1	1
3	3	1	−2	−3	9
4	0	3	3	2	4
5	6	5	−1	−2	4
Total			5		22

$$\bar{d} = \frac{5}{5} = 1.0$$

$$s_{\bar{d}} = \sqrt{\frac{22}{4}} = \sqrt{5.5}$$

$$t = \frac{1}{\sqrt{5.5}/\sqrt{5}} = \frac{1}{\sqrt{1.1}} = 0.953, \text{ with df} = (5-1) = 4$$

Since, the calculated value is less than the table value of $t(4) = 2.78$ at 5 percent level of significance, the null hypothesis is accepted ($P > 0.05$, not significant).

KEYWORDS

- Null hypothesis
- Parametric tests
- t-test

CHAPTER 10

EXPERIMENTAL DATA ANALYSIS: ANOVA

CONTENTS

10.1   ONE-WAY ANOVA

10.1.1   LAYOUT

The one-way ANOVA is carried out for data in completely randomized design (CRD). The layout of CRD is as given below:

	Treatment 1	Treatment 2	-	Treatment k
	x_{11}	x_{21}		x_{k1}
	x_{12}	x_{22}		x_{k2}
	x_{1n_1}	x_{2n_2}		x_{kn_k}
Mean	$\overline{x}_1$	$\overline{x}_2$	-	$\overline{x}_k$

10.1.2   HYPOTHESES

The one-way ANOVA is applied to test for homogeneity of several population means. The null hypothesis states that the population means are homogenous (equal). That is,

$$H_0: \mu_1 = \mu_2 = \ldots = \mu_k \text{ against } H_1: \mu_i\text{'s are not equal.}$$

10.1.3   TECHNIQUE

The technique consists in estimating two population variances: one is based on between-sample variance called the mean sum of squares between groups or samples (MSS_B) and the other is based on within-sample variance called mean sum of squares within samples (MSS_w). Then, the two estimators are compared with the F-ratio. When the samples come from identical population, these two estimates of the population variances are comparable; any difference observed between them is expected to be within the range of their sampling error. When the populations from which the samples are drawn have different means, the MSS_B is expected to provide a higher value. Symbolically, the technique of ANOVA is given as follows:

$$SS_T = SS_B + SS_W$$

10.1.4   TEST STATISTIC

The test statistic is given by

$$F = \frac{MSS_B}{MSS_W} \text{ with df} = (k - 1, n - k)$$

ANOVA table: It is given by

Sources of Variation	df	SS	MSS	F-Ratio
Between treatment	$(k-1)$	SS_B	MSS_B	F
Within treatment	$(n-k)$	SS_W	MSS_W	-
Total	$(n-1)$	SS_T	-	-

where $MSS_B = \dfrac{SS_B}{(k-1)}$

$$MSS_W = \frac{SS_W}{(n-k)}$$

$$SS_B = \sum n_i \, (\overline{x}_i - \overline{x})^2 = \sum \frac{T_i^2}{n_i} - \frac{T^2}{n}$$

$$SS_W = \sum\sum (X_{ij} - \overline{x}_i)^2 = \sum\sum x_{ij}^2 - \sum \frac{T_i^2}{n_i}$$

Example: Let us suppose that the improvement scores of 15 patients from three different treatments are given below:

Treatment 1	Treatment 2	Treatment 3
4	7	10
2	4	7
3	1	8
0	3	6
6	5	4

For this data, the one-way ANOVA computations are as given below:

Statistics	Treatment 1	Treatment 2	Treatment 3
Sample size	5	5	5
Sum	15	20	35
Mean	3	4	7
Sum of squares	65	100	265

Hence,

$$SS_B = \left(\frac{15^2}{5} + \frac{20^2}{5} + \frac{35^2}{5} \right) - \frac{70^2}{15} = 370 - 326.67 = 43.33$$

$$SS_W = (65 + 100 + 265) - 370 = 430 - 370 = 60$$

Source of Variation	df	SS	MSS	F-ratio
Between treatment	2	43.33	21.67	4.33
Within treatment	12	60.00	5.00	-
Total	14	103.33	-	-

Since the calculated value of F is more than the table value of F (2, 12) = 3.89, we reject the null hypothesis at 5 percent level of significance ($P < 0.05$, significant).

10.2   POST-HOC MULTIPLE COMPARISON TESTS: Scheffé's METHOD

10.2.1   HYPOTHESES

A significance F-test in ANOVA does not mean that every group in the analysis is significantly different from every other group. Sometimes, we are interested in carrying out the test of significance of the difference in pairs of means. The tests applied after finding out that the overall F is significant are called post-hoc tests. Here, the usual Z-values and t-values are cumbersome. The Scheffé's method is the commonly used post-hoc test. This method consists in calculating critical value of F (denoted by F_{cv}) to use as a standard against which to compare the differences in pairs of means.

10.2.2   CRITICAL VALUE OF F

It is given by

$$F_{cv} = (k - 1) F$$

   where k is the number of groups
   F is the value needed to gain significance in the ANOVA

10.2.3   TEST STATISTICS

The test statistics for equality of means of ith group and the jth group in this method is given by

$$F_{(i,j)} = \frac{(\bar{x}_i - \bar{x}_j)^2}{MSS_w \left(\dfrac{1}{n_i} + \dfrac{1}{n_j} \right)}$$

The F is the value needed to gain significance in the ANOVA.

   Example: Let us apply the Scheffé's method of post-hoc multiple comparison test to compare pairs of means to the data used to demonstrate the one-way ANOVA.

   Here,

$$F_{cv} = (3 - 1) \; 3.89 = 7.78$$

To compare group 1 and group 2,

$$F_{(1,2)} = \frac{(3-4)^2}{5 \left(\dfrac{1}{5} + \dfrac{1}{5} \right)} = \frac{1}{2} = 0.5$$

To compare group 1 and group 3,

$$F_{(1,3)} = \frac{(3-7)^2}{5 \left(\dfrac{1}{5} + \dfrac{1}{5} \right)} = \frac{16}{2} = 8.0$$

To compare group 2 and group 3,

$$F_{(2,3)} = \frac{(4-7)^2}{5\,(\frac{1}{5}+\frac{1}{5})} = \frac{9}{2} = 4.5$$

Since the *Scheffé* $F_{(1,3)}$ is more than F_{cv}, we reject the null hypothesis that the means of first group and the third group are different ($P < 0.05$, significant).

10.3   TWO-WAY ANOVA

10.3.1   LAYOUT

The two-way ANOVA is carried out for data in randomized block design (RBD). The layout of the data collected with this design is as given below:

Blocks	Treatment 1	Treatment 2	-	Treatment k	Mean
1	x_{11}	x_{21}	-	x_{k1}	$\bar{x}.1$
2	x_{12}	x_{22}	-	x_{k2}	$\bar{x}.2$
-	-	-	-	-	-
r	x_{1r}	x_{2r}	-	x_{kr}	$\bar{x}.r$
Mean	$\bar{x}_1 .$	$\bar{x}_2 .$	-	$\bar{x}_k .$	$\bar{x}$

10.3.2   HYPOTHESES

To test the significance of the means of t treatments effects, and also the significance of the means of r blocks effects. That is,

$H_{01}: t_1 = t_2 = - - = t_k$ against H_{11}: The treatments have different effects
$H_{02}: b_1 = b_2 = - - = b_r$ against H_{12}: The blocks have different effects

10.3.3   TECHNIQUE

The technique consists in dividing the total variance as given below:
$SS_T = SS_{Tr} + SS_{Bl} + SS_E$

10.3.4   TEST STATISTICS

They are given by

$$F_{Tr} = \frac{MSS_{Tr}}{MSS_E} \text{ with df} = (k-1), (r-1)(k-1)$$

$$F_{BL} = \frac{MSS_{BL}}{MSS_E} \text{ with df} = (r-1), (r-1)(k-1)$$

Two-Way ANOVA Table

Source of Variation	df	SS	MSS	F-ratio
Between treatments	$(k-1)$	SS_{Tr}	MSS_{Tr}	F_{Tr}
Between blocks	$(r-1)$	SS_{Bl}	MSS_{Bl}	F_{Bl}
Error	$(r-1)(k-1)$	SS_E	MSS_E	-
Total	$(rk-1)$	SS_T	-	-

Example: Let us support that the improvement scores of 15 patients from three treatments and five (groups) blocks using RBD is given below:

Blocks	Treatment 1	Treatment 2	Treatment 3	Total	Mean
1	4	7	10	21	7.0
2	2	4	7	13	4.3
3	3	1	8	12	4.0
4	0	3	6	9	3.0
5	6	5	4	15	5.0
Mean	3	4	7	70	4.6

The two-way ANOVA is given by

Source of Variation	df	SS	MSS	F-ratio
Between treatment	2	43.33	21.67	5.20
Between blocks	4	26.67	6.67	1.60
Error	8	33.33	4.17	-
Total	14	103.33	-	-

The requisite computations are

$$SS_{Bl} = \frac{1}{3}(21^2 + 13^2 + 12^2 + 9^2 + 15^2) - 326.67$$

$$= \frac{1060}{3} - 326.67 = 353.33 - 326.67 = 26.67$$

Since, the calculated value of F_{Tr} is more than the table value of $F(2,8)$ = 4.46, the null hypothesis is rejected ($P < 0.05$, significant) in cases of treatments.

10.4   2²-FACTORIAL ANOVA

10.4.1   LAYOUT

The layout of the data in 2^2 (or 2×2) factorial experimental design with RBD is given below:

Blocks	1	a	b	ab
1	x_{11}	x_{21}	x_{31}	x_{41}
2	x_{12}	x_{22}	x_{32}	x_{42}
-	-	-	-	-
R	x_{1r}	x_{2r}	x_{3r}	x_{4r}
Total	(1)	(a)	(b)	(ab)

10.4.2   HYPOTHESES

The four treatment combinations can be compared by carry out a 2²-factorial ANOVA.

10.4.3   TECHNIQUE

In 2²-factorial ANOVA, our main objective is to carry out separate test for the main effects A, B, and the interaction AB by splitting the SS_{Tr} with 3 df in three orthogonal components each associated with either with the main effects A and B or the interaction AB, and each with one degree of freedom. That is,

$$SS_{Tr} = SS_A + SS_B + SS_{AB}$$

10.4.4   TEST STATISTICS

The test statistics are given by

$F_A = \frac{MSS_A}{MSS_E}$ with df = 1, 3(r − 1)

$F_B = \frac{MSS_B}{MSS_E}$ with df = 1, 3(r − 1)

$F_{AB} = \frac{MSS_{AB}}{MSS_E}$ with df = 1, 3(r − 1)

where $SS_A = \left(-(1)+(a)\ -(b)+(ab)\right)^2/4r$

$SS_B = \left(-(1)-(a)\ +(b)+(ab)\right)^2/4r$

$SS_{AB} = \left((1)-(a)\ -(b)+(ab)\right)^2/4r$

2^2-Factorial ANOVA Table:

Source of Variation	df	SS	MSS	F-ratio
Treatments	3	SS_{Tr}	MSS_{Tr}	F_{Tr}
Main A	1	SS_A	MSS_A	F_A
Main B	1	SS_B	MSS_B	F_B
Interaction AB	1	SS_{AB}	MSS_{AB}	F_{AB}
Blocks	(r-1)	SS_{BL}	MSS_{Bl}	F_{Bl}
Errors	3 (r-1)	SS_E	MSS_E	-
Total	(4r-1)	SS_T	-	-

Example: Let us support that there are two different treatments A and B each two levels (not given, given), and hence there are four groups of patients each group consisting of five patients allotted at random to five age groups (2^2-factorial design in RBD). The improvement scores of the 20 patients are given below:

Blocks	Treatments				Mean
	1	a	b	ab	
1	4	7	10	9	7.50
2	2	4	7	12	6.25
3	3	1	8	8	5.00
4	0	3	6	10	4.75
5	6	5	4	6	5.25
Mean	3	4	7	9	5.75

The 2^2-factorial ANOVA table is given below:

Score of Variation	df	SS	MSS	F-ratio
Treatment	3	113.75	37.92	7.74
Main A	1	11.25	11.25	2.30
Main B	1	101.25	101.25	20.67
Interaction AB	1	1.25	1.25	0.67
Blocks	4	20.50	5.13	1.05
Errors	12	59.50	4.90	-
Total	19	193.75	-	-

Since the calculated value is more than the table value of $F(3, 12) = 3.49$, the treatments effects are significantly different ($P < 0.05$, significant).

10.5   REPEATED MEASURES ANOVA

10.5.1   LAYOUT

The layout of repeated measures ANOVA design is given by

Subject	Repeat 1	Repeat 2	-	Repeat k
1	x_{11}	x_{21}	-	x_{k1}
2	x_{12}	x_{22}	-	x_{k2}
-	-	-	-	-
n	x_{1n}	x_{2n}	-	x_{kn}

10.5.2   HYPOTHESES

To test homogeneity of means of several repeated measures.

10.5.3   TECHNIQUE

Here, we are interested in refining the experimental design to increase its sensitivity to detect differences in the dependent variable. A major source of experimental error is individual differences that can be controlled by using repeated measures or within-subjects design. By having the same participants perform under every condition, systematic bias attributable to participants in one group being different from the participants in other groups can be eliminated. The assumptions underlying this method are random selection, normality, homogeneity of variance, and sphericity. In this technique,

$$SS_{Rep} = SS_B$$

$$SS_{Res} = SS_W - SS_{Sub}$$

where

$$SS_{Sub} = \sum \frac{T_{.k}^2}{k} - \frac{T^2}{nk}$$

10.5.4   TEST STATISTIC

It is given by

$$F_{Res} = \frac{MSS_{Rep}}{MSS_{Res}}$$

where $SS_{Sub} = \sum \frac{T_{.k}^2}{k} - \frac{T^2}{nk}$

Repeated Measures ANOVA Table:

Source of Variation	df	SS	MSS	F-ratio
Repeated measures	(k–1)	SS_{Rep}	MSS_{Rep}	F_{Rep}
Residuals	(n–1) (k–1)	SS_{Res}	MSS_{Res}	-

Example: Let us suppose that the marks scored by five students in three repeated tests are given below:

Student	Repeat 1	Repeat 2	Repeat 3
1	4	7	10
2	2	4	7
3	3	1	8
4	0	3	6
5	6	5	4

For this data, the repeated measures ANOVA is given below:

Source of Variation	df	SS	MSS	F-ratio
Repeated measures	2	43.33	21.67	5.20
Residuals	8	33.33	4.17	-

Since, the calculated value of F is more than the table value of $F(2, 8)$ = 4.46, we reject the null hypothesis ($P < 0.05$, significant)

KEYWORDS

- ANOVA
- Randomized block design
- Scheffé's method
- Test statistics

CHAPTER 11

NON-PARAMETRIC TESTS OF SIGNIFICANCE

CONTENTS

11.1   ONE-SAMPLE NON-PARAMETRIC TESTS

11.1.1   *CHI-SQUARE TESTS: 1x K CONTINGENCY TABLE*

11.1.1.1   *HYPOTHESES*

The one-sample chi-square test may be used to test for goodness of fit. It can be applied to decide whether a distribution of a variable in a sample represents or fits a specified population. That is, you can use this test to decide whether your data follows approximate normal distribution, uniform distribution, Poisson distribution, etc.

11.1.1.2   *1x K CONTINGENCY TABLE*

A table showing bivariate frequency distribution is known as contingency table. The 1×k contingency table has one row and k columns as shown below:

1 x k Contingency Table

Attribute	Category 1	Category2	-	Category k	Total
Observed Frequencies	O_1	O_2	-	O_k	n

11.1.1.3   *TEST STATISTIC*

It is given by

$$\chi^2 = \sum \frac{(O_i - e_i)^2}{e_i} = \sum \frac{O_i^2}{e_i} - n \text{ with df} = (k-1)$$

where O_i is the observed frequency of the ith category; e_i is the expected frequency of the ith category. They are calculated under the assumption that the null hypothesis is true.

Example: The classification of the 40 registered patients (data in Appendix I) according to the type of locality is given in the following table. We wish to test whether the number of registrations from the three localities can be considered equal.

	Rural	Semi-urban	Urban	Total
Number of patients	16	13	11	40

For this data,

	Rural	Semi-urban	Urban	Total
O_i	16	13	11	40
e_i	13.33	13.33	13.33	40
$(O_i - e_i)$	2.67	−0.33	−2.33	0
$(O_i - e_i)^2$	7.13	0.11	5.43	
$\dfrac{(O_i - e_i)^2}{e_i}$	0.53	0.007	0.398	0.953

Hence,
$\chi^2 = 0.953$, df $= (3-1) = 2$

Since, the calculated value is less than the table value of $\chi^2 = 5.99$ at df $= 2$, we accept the null hypothesis ($P > 0.05$, not significant).

11.1.2   CHI-SQUARE TESTS: *R* x *K* CONTINGENCY TABLE

11.1.2.1   HYPOTHESES

The chi-square test may be used to test for independence of attributes.
   That is,
   H_0: The two attributes are independent.

11.1.2.2   R x K CONTINGENCY TABLE

The $r \times k$ contingency table has r rows and c columns as shown below:

r × k Contingency Table

Variable 2 / Variable 1	Category 1	Category 2	-	Category k	Total
Category 1	O_{11}	O_{12}		O_{1k}	O_1
Category 2	O_{21}	O_{22}		O_{2k}	O_2
-					
Category r	O_{r1}	O_{r2}		O_{rk}	O_r
Total	$O_{.1}$	$O_{.2}$		$O_{.k}$	n

11.1.2.3   TEST STATISTIC

It is given by,

$$\chi^2 = \sum \frac{(O_{ij} - e_{ij})^2}{e_{ij}} = \sum \frac{O_{ij}^2}{e_{ij}} - n \text{ with df} = (r-1)(c-1)$$

where

$$e_{ij} = \frac{O_i O_j}{n}$$

Example: A classification according to religion and diagnostic categories of 337 psychiatric patients is given in the following table (figures in the bracket are expected frequencies).

Disorder/ Religion	Schizophrenia	Affective Disorder	Neurotic Disorder	Total
Hindu	109 (102.7)	114 (115.1)	56 (61.3)	279
Muslim	8 (16.2)	22 (18.1)	14 (9.7)	44
Christian	7 (5.2)	3 (5.8)	4 (3.1)	14
Total	124	139	74	337

For this data,

$$\chi^2 = \left(\frac{109^2}{102.7} + \frac{114^2}{115.1} \cdots + \frac{4^2}{3.1} \right) - 337 = 347.034 - 337 = 10.03$$

$$\text{df} = (3-1)(3-1) = 4$$

The expected frequencies are calculated based on the null hypothesis that the two attributes are independent. For example, the expected frequency for Hindu schizophrenia is calculated as follows:

$$e_{11} = (279 \times 124)/337 = 102.7$$

Since the calculated value is more than the table value of $\chi^2 = 9.49$ at df $= 4$, we reject the null hypothesis ($P < 0.05$, significant).

11.1.3 CHI-SQUARE TESTS: 2x2 CONTINGENCY TABLE

11.1.3.1   HYPOTHESES

The chi-square can be used to test whether two binary attributes are independent or not.

11.1.3.2   2 x 2 CONTINGENCY TABLE

The 2×2 contingency table has two rows and two columns as shown below:

2×2 Contingency Table

Variable 2/ Variable 1	Category 1	Category 2	Total
Category 1	a	b	(a + b)
Category 2	c	d	(c + d)
Total	(a + c)	(b + d)	n

11.1.3.3   TEST STATISTIC

It is given by

$$\chi^2 = \frac{n(ad-bc)^2}{(a+b)(c+d)(a+c)(b+d)} \text{ with df} = 1$$

Example: The classification according to sex and marital status of the 40 patients (data in Appendix I) is as given below:

Marital Status/Sex	Single	Married	Total
Male	14	10	24
Female	7	9	16
Total	21	19	40

For this data,

$$\chi^2 = \frac{40\,(14 \times 9 - 10 \times 7)^2}{24 \times 16 \times 21 \times 19} = \frac{40\,(126 - 70)^2}{24 \times 16 \times 21 \times 19} = 0.82, \text{ with df} = 1$$

Since the calculated value is less than the table value of $\chi^2 = 3.841$, with df = 1, we accept the null hypothesis.

11.1.4   CHI-SQUARE TESTS: 2xK CONTINGENCY TABLE

11.1.4.1   HYPOTHESES

The chi-square test can be used to test whether the two attributes are independent or not.

11.1.4.2   2xK CONTINGENCY TABLE

A 2×k contingency table has two rows and k columns as shown below:

2×k Contingency Table

Variable 2/Variable 1	Category 1	Category 2	-	Category k	Total
Category 1	X_1	X_2		X_k	X
Category 2	Y_1	Y_2		Y_k	Y
Total	Z_1	Z_2		Z_k	Z

11.1.4.3   TEST STATISTIC

It is given by

$$\chi^2 = \frac{Z^2}{XY}\left(\frac{X_1^2}{Z_1} + \frac{X_2^2}{Z_2} + \ldots + \frac{X_k^2}{Z_k}\right) - \frac{ZX}{Y} \text{ with df} = (k-1)$$

Example: A classification according to type of service and religion of 500 psychiatric patients registered at NIMHANS hospital is as presented below:

Religion/Admission Status	Hindu	Muslim	Christian	Total
In-patients	133(32.8)	14(21.2)	11(39.3)	158(31.6)
Out-patients	273(67.2)	52(78.8)	17(60.7)	342(68.4)
Total	406(100)	66(100)	28(100)	500(100)

For this data,

$$\chi^2 = \frac{500^2}{158 \times 342}\left(\frac{133^2}{406} + \frac{14^2}{66} + \frac{11^2}{28}\right) - \frac{500 \times 158}{342} = 4.39 \text{ with df} = (k-1)$$

Since the calculated value is less than the table value of chi-square at 2 df (5.99), we accept the null hypothesis ($P > 0.05$, not significant).

11.1.5   TESTS OF SIGNIFICANCE OF INDIVIDUAL CELL FREQUENCIES

11.1.5.1   HYPOTHESES

This is the identification of significantly high/low frequencies of individual cells in a contingency table.

11.1.5.2   CONTINGENCY TABLE

The tests of significance of individual cell frequencies are carried out for contingency table of any dimension.

11.1.5.3   TEST STATISTICS

It consists in computing adjusted standardized residuals (Z_{ij}) for each cell frequency in the contingency table. Thus, the test statistics are given by

$$Z_{ij} = \frac{\epsilon_{ij}}{\sqrt{\text{var}(\epsilon_{ij})}}$$

where the standardized residuals (ϵ_{ij}) are given by

$$\epsilon_{ij} = \frac{(o_{ij} - e_{ij})}{\sqrt{e_{ij}}}$$

$$\text{Var}(\epsilon_{ij}) = (1 - p_i)(1 - p_j)$$

where p_i and p_j are the proportions of the marginal totals of the ij-th cell. The ϵ_{ij} are called standardized residuals of the ij-th cell. When the variables forming the contingency table are independent, the Z_{ij} are approximately normally distributed with mean 0 and variance 1.

Example: The observed frequencies and the expected frequencies (in the brackets) of the 40 patients (data in Appendix I) classified according to sex and marital status is given below:

Sex	Single	Married	Total
Male	14 (12.6)	10 (11.4)	24
Female	7 (8.4)	9 (7.6)	16
Total	21	19	40

The residuals $(O_{ij} - e_{ij})$ are given in the following table.

Sex	Single	Married	Total
Male	1.4	−1.4	0
Female	−1.4	1.4	0
Total	0	0	0

The standardized residuals are calculated and presented in the following table.

Sex	Single	Married
Male	0.39	−0.41
Female	−0.48	0.51

Finally, the adjusted standardized residuals Z_{ij} are computed as shown in the following table:

Sex	Single	Married
Male	0.9	−0.9
Female	−0.9	0.9

11.1.6   ONE-SAMPLE RUN TEST

11.1.6.1   HYPOTHESES

The run test is applied to test whether the sequence of events in a sample is at random. Thus, it is a randomization test.

11.1.6.2   SEQUENCE OF EVENTS

The test is based on the number of runs which a sample exhibits, the number of objects in one category (n_1), and the number of objects in another category (n_2). A run is defined as a success of identical symbols that are followed and proceed by different symbols or no symbol at all.

11.1.6.3   TEST STATISTIC

The test statistic is given by

$$Z = \frac{|r - \mu_r|}{\sigma_r}$$

where $\mu_r = \frac{2n_1 n_2}{n_1 + n_2} + 1$

$$\sigma_r = \sqrt{\frac{2n_1 n_2 (2n_1 n_2 - n_1 - n_2)}{(n_1 + n_2)^2 (n_1 + n_2 - 1)}}$$

Too many or two small number of runs do not support randomness.

Example: The 40 patients (data in Appendix I) were registered in the following order of males (M) and females (F):

M FFF MM F MM F M FFF MM F MMMMMMMMM FFF MM F MMMM F M FF

For this data,
   Number of males (n_1) = 24
   Number of females (n_2) = 16
   Number of runs (r) = 18
   Hence,

$$\mu_r = \frac{2 \times 24 \times 16}{24 \quad 16} + 1 = 19.2 + 1 = 20.2$$

$$\sigma_r = \sqrt{\frac{(2 \times 24 \times 16)(2 \times 24 \times 16 - 24 - 16)}{(24+16)^2 (24+16-1)}} = \sqrt{8.96} = 2.993$$

$$Z = \frac{|18 - 20.2|}{2.993} = \frac{2.2}{2.993} = 0.735$$

Since the calculated value is less than 1.96, we accept the null hypotheses that the sequence of occurrence of males and females is at random.

11.2 TWO INDEPENDENT SAMPLE NON-PARAMETRIC TESTS

11.2.1 MEDIAN TEST

The median test may be used when the null hypothesis states that the two groups are from population with the same median. This method is applicable whenever the scores of the two groups are in at least an ordinal scale. To perform the median test, we first determine the median score of the combined group. Then, we dichotomize both sets of scores at that combined median. Then, the procedure is to prepare a 2×2 contingency table and apply the chi-square test of significance.

11.2.2   MANN–WHITNEY U-TEST

11.2.2.1   HYPOTHESES

The Mann–Whitney U-test is used to test whether two independent groups have been drawn from the same population. Thus, it is equivalent to the independent sample t-test.

11.2.2.2   MANN–WHITNEY U-TEST TABLE

The method consists in combining the two samples and arranges the observations in order of magnitude. Then, assign ranks to the combined observations from 1 to $(n_1 + n_2)$.

11.2.2.3   TEST STATISTIC

Now, the test statistic is given by

$$Z = \frac{|U - \mu_U|}{\sigma_u}$$

where $U = n_1 n_2 + \dfrac{n_1(n_1 + 1)}{2} - R_1$

R_1 is the sum of the ranks of the first group.

$$\mu_u = \frac{n_1 n_2}{2}$$

$$\sigma_u = \sqrt{\frac{n_1 n_2 (n_1 + n_2 + 1)}{12}}$$

Example: Let us suppose that the marks scored in a test by ten randomly selected students from two departments are given below. We wish to test whether the marks scored by the students in the two departments are the same.

Department 1	Department 2
4	9
2	7
3	8
0	5
6	11

For this data, the scores are converted as ranks as shown below:

Department 1	Department 2
4	9
2	7
3	8
1	5
6	10
16	39

Here,

$R_1 = 16$

$U = (5 \times 5) + \dfrac{5(5+1)}{2} - 16 = 25 + 15 - 16 = 24$

$\mu_u = \dfrac{5 \times 5}{2} = 12.5$

$\sigma_u = \sqrt{\dfrac{5 \times 5 \ (5+5+1)}{12}} = 4.787$

Hence,

$Z = \dfrac{24-12.5}{4.787} = \dfrac{11.5}{4.787} = 2.402$

Since, the calculated value is more than 1.96, we reject the null hypotheses of randomness ($P < 0.05$, significant).

11.3   TWO-RELATED SAMPLE NON-PARAMETRIC TESTS

11.3.1   McNEMAR'S TEST

11.3.1.1   HYPOTHESES

The McNemar's test is used when the researcher involved in hypotheses that there is change after the treatment. This test of significance is used when both the variables under study are measured in nominal scale.

11.3.1.2   McNEMAR'S TEST TABLE

The researcher sets up a four-fold table of frequencies to represent first and second sets of responses from the same individual. A typical 2×2 McNemar's test table is given below, where fail and pass are used to signify different responses.

2×2 McNemar's test table

		After	
		Pass	**Fail**
Before	fail	A	B
	pass	C	D

11.3.1.3   TEST STATISTIC

Now, the test statistic is given by

$$\chi^2 = \frac{(A - D)^2}{(A + D)} \text{ with df} = 1.$$

Example: Let us suppose that the results of a test of 36 students before giving the teaching program and the results after taking the teaching program are given in the following table. We wish to test whether there is a significant effect of the training program.

		After		Total
		Pass	Fail	
Before	Fail	16	2	18
	Pass	14	4	18
	Total	30	6	36

$$\chi^2 = (16 - 4)^2/(16 + 4) = 144/20 = 7.20, \text{ with df} = 1$$

Since the calculated value is more than the table value of $\chi^2 = 6.63$, with df $= 1$ at 1 percent level of significance, we reject the null hypotheses ($P < 0.01$, highly significant). Hence, there is a significant effect of the teaching program.

11.3.2   SIGN TEST

11.3.2.1   HYPOTHESES

The sign test is used when the researcher involves in hypothesis that there is change after the treatment. This test is applied when the data on a variable have underlying continuity but which can be measured in only a very gross way.

11.3.2.2   SIGN TEST TABLE

The sign test uses plus and minus signs rather than quantitative measures as its data. The method consists in focusing on the direction of the difference between every pair of the observations and record plus sign when the other values is less than the value before giving the treatment, and minus sign otherwise.

11.3.2.3   TEST STATISTIC

Now, the test statistic is given by

$$Z = \frac{|x - 0.5n|}{0.5\sqrt{n}}$$

where n is the number of observations

x is the number of plus signs.

Example: Let us suppose that the results of 36 students before taking the teaching program and the results of the students after taking the teaching program are given: it is found that 31 students have improved. We wish to test whether there is a significant effect (improvement) of the teaching program.

$$Z = \frac{|31 - 0.5 \times 36|}{0.5\sqrt{36}} = \frac{13}{3} = 4.33$$

Since the calculated value is more than 2.58, the null hypothesis is rejected ($P < 0.01$, highly significant). That is, there is a significant effect of the teaching program.

11.4   K-INDEPENDENT SAMPLE NON-PARAMETRIC TESTS

11.4.1   *EXTENDED MEDIAN TEST*

This is an extension of the median test in which there are more than two groups in the data for this test, and the researcher is interested to test whether these groups are from populations with the same median. The requisite 2×k contingency table has to be prepared, and the chi-square test of significance has to be applied.

11.4.2   KRUSHKAL–WALLIS *H*-TEST

11.4.2.1   *HYPOTHESES*

This test that is based on ranks is to test whether the k-samples are from the same population.

11.4.2.2   KRUSHKAL–WALLIS H-TEST TABLE

In this test, each of the n observations is replaced by ranks. That is, all the scores from all of the k-samples combined are ranked in a single series. The smallest score is replaced by rank 1, the next to smallest by rank 2, and the largest score by rank n. Here, n is the total number of independent observations in the k-samples. Then, the sum of the ranks in each sample (column) is found. To test whether these sum of ranks are so desperate that they are not likely to have come from samples that were all drawn from the same population.

11.4.2.3   TEST STATISTIC

Now, the test statistic is given by

$$H = \frac{12}{n(n+1)} \sum \frac{R_j^2}{n_j} - 3\,(n+1)$$

The H follows approximate chi-square with df $= (k - 1)$
where k is the number of samples (groups)
n is the number of cases in all the samples combined
n_j is the number of cases in the jth sample
R_j is the sum of the ranks in the jth sample
Example: Let us suppose that the marks scored by 15 randomly selected students from three departments are as given in the following table.

Department 1	Department 2	Department 3
4	9	14
2	7	12
3	8	13
0	5	10
6	11	16

We wish to test whether the scores from the three departments are same. The ranks of the observations (scores) in the three departments combined are given in the following table.

	Department 1	Department 2	Department 3
	4	9	14
	2	7	12
	3	8	13
	1	5	10
	6	11	15
Sum of ranks	16	40	64
Mean ranks	3.2	8.0	12.8

For this data,

$$H= \frac{12}{(15 \times 16)}\left(\frac{16^2}{5}+\frac{40^2}{5}+\frac{64^2}{5}\right)-3 \times 6 = 59.52 - 48 = 11.52$$

with df $= (3-1)=2$

Since the calculated value is more than the table value of $\chi^2 (2) = 9.21$, we reject the null hypothesis ($P < 0.01$, highly significant).

11.5   K-RELATED SAMPLE NON-PARAMETRIC TESTS

11.5.1   COCHRAN Q-TEST

11.5.1.1   HYPOTHESES

This is an extension of the McNemar's test in which there are more than two related samples for Cochran Q-test. This provides a method for testing whether three or more matched sets of frequencies or proportion differ significantly among themselves. This test is particularly suitable when the data are dichotomized ordinal information.

11.5.1.2   COCHRAN Q-TEST TABLE

A typical Cochran Q-test table is as given below:

Cochran Q-test Table

Serial Number	Repeat 1	Repeat 2	Repeat k	Total
1				L_1
2				L_2
-				-
n				L_n
Total (G)	G_1	G_2	G_k	

11.5.1.3   TEST STATISTIC

A typical $n \times k$ Cochran Q-test table is given below (pass, 1; fail, 0) in which students are tested in three occasions.

Student	Repeat 1	Repeat 2	Repeat 3
1	1	1	1
2	0	1	1
3	0	0	1
4	0	0	1
5	1	1	1

Now, the test statistic is given by

$$Q = \frac{(k-1)\left(k\sum G_j^2 - (G_j)^2\right)}{k\sum L_i - \sum L_i^2}$$

The Q follows approximate χ^2 with df $= (k-1)$

where k is the number of repetitions (conditions)

G_j is the sum of the jth column

L_i is the sum of the ith row

Example: Let us suppose that the scores obtained on three repetitions of five students are as given in the following table.

Student	Repeat 1	Repeat 2	Repeat 3	L_i	L_i^2
1	1	1	1	3	9
2	0	1	1	2	4
3	0	0	1	1	1
4	0	0	1	1	1
5	1	1	1	3	9
G_j	2	3	5	10	24
G_j^2	4	9	25		38

$$Q = \frac{2(3 \times 38 - 10^2)}{(3 \times 10 - 24)} = \frac{2(114 - 100)}{(30 - 24)} = \frac{28}{6} = 4.67 \text{ with df} = (3-1) = 2$$

Since the calculated value is less than the Table value of $\chi^2 = 5.99$ with df $= 2$, we accept the null hypotheses ($P > 0.05$, not significant). That is, the teaching program has no significant effect.

11.5.2 FRIEDMAN TEST

11.5.2.1 HYPOTHESES

The Friedman test of significance is used when the researcher has the hypothesis that the k-related sample has come from different populations with respect to mean ranks. The test is applicable when the measurement of the variable are expressed as ranks.

11.5.2.2 FRIEDMAN TEST TABLE

It is given by,
Friedman test table

Serial Number	Rank (repeat 1)	Rank (repeat 2)	-	Rank (repeat k)
1	-	-	-	-
2	-	-	-	--
-	-	-	-	-
n	-	-	-	-
Sum	R_1	R_2	-	R_k

The data are given in a two-way table having n rows and k columns. The rows represent the various subjects or matched set of subjects and the columns represent the various conditions. The data of the test are ranks. The scores in each row are ranked separately with k conditions being studied, the ranks in any row ranked from 1 to k.

11.5.2.3   TEST STATISTIC

Now, the test statistic is given by

$$\chi^2 = \frac{12}{nk(k+1)} \sum R_j^2 - 3n\,(k+1) \text{ with df} = (k-1)$$

where n is the number of rows (patients, subjects)

k is the number of columns (conditions)

R_j is the sum of ranks of the jth column

Example: Let us suppose that the marks scored by five students (on a teaching program) in three consecutively conducted tests are given below. We wish to test whether there is a significant improvement of teaching program in repeated tests.

Student	Repeat 1	Repeat 2	Repeat 3
1	4	7	10
2	2	4	7
3	3	1	8
4	0	3	6
5	6	5	4

The requisite computation table (ranks of the marks) is as given below:

Student	Rank (repeat 1)	Rank (repeat 2)	Rank (repeat 3)
1	1	2	3
2	1	2	3
3	2	1	3
4	1	2	3
5	3	2	1
Sum	8	9	13
Mean	1.6	1.8	3.6

Now,

$$\chi^2 = \frac{12}{(5 \times 3 \times 4)}(8^2 + 9^2 + 13^2) - (3 \times 5 \times 4) = \frac{1}{5}(64 + 81 + 169) - 60 = 2.80,$$

with df $= (3 - 1) = 2$

Since, the calculated value is less than the table value of $\chi^2 = 5.99$, with df $= 2$, we accept the null hypothesis ($P > 0.05$, not significant). That is, there is no significant improvement in the repeated administration of the program.

11.6 LOG-LINEAR MODELS

11.6.1 INTRODUCTION

The term "model" refers to some theory or conceptual framework about the observations. The parameters in the model represent the effects that particular variables or combinations of variables have in determining the observed values. Such an approach is common in regression analysis and analysis of variance (ANOVA). Most common are linear models, which postulate that the expected values of the observations are given by a linear combination of a number of parameters. Techniques such as maximum likelihood and least squares may be used in estimating the parameters. Then the estimated parameter values are used in identifying variables that have the greatest importance in determining the observed values. The

ANOVA term "interactions" is used instead of the term "association" for describing a relationship between the qualitative variables forming a contingency table. We shall speak of first-order interactions between pairs of variables, second-order interactions between triplets of variables, and so on. The major advantages of these techniques are that they provide a systematic approach to the analysis of complex multidimensional tables, provide estimates of the magnitude of effects of interest, and consequently they allow the relative importance of different effects to be judged.

11.6.2 TWO-DIMENSIONAL TABLES

Let us consider how the type of model used in the analysis of variance of quantitative data can arise for contingency table data. Let us deal with the two-dimensional table and the hypothesis of independence. That is no first-order interaction between the two variables. It is specified by,

$$P_{ij} = P_{i.} \, P_{.j}$$

This relationship specifies a model for the data. It is that in the population the probability of an observation falling in the ij^{th} cell of the table is simply the product of the marginal probabilities. We wish to ask how this model could be rearranged so that P_{ij} can be expressed as the sum of the marginal probabilities or some function of them. By taking the natural logarithms of the above expression, we found,

$$\ln P_{ij} = \ln P_{i.} + \ln P_{.j}$$

$$e_{ij} = nP_{ij} = n \, P_{i.} \, P_{.j}$$

$$= n \, \frac{e_{i.} e_{.j}}{nn} = \frac{e_{i.} e_{.j}}{n}$$

We have, $\ln e_{ij} = \ln e_{i.} + \ln e_{.j} - \ln n$

This equation may be rewritten in a form reminiscent of the models used in ANOVA namely:

$$\ln e_{ij} = u + u_1(i) + u_2(j)$$

This is the linear model for the logarithms of the frequencies or what is generally known as log-linear model. Here,

$$u = \frac{\sum\sum \ln e_{ij}}{rc}$$

$$u_1(i) = \frac{\sum_{j=1}^{c} \ln e_{ij}}{c} - u$$

$$u_2(j) = \frac{\sum_{i=1}^{r} \ln e_{ij}}{r} - u$$

Where u is the overall mean effect, $u_1(i)$ is the main effect of the ith category of variable 1 and $u_2(j)$ is the main effect of the jth category of variable 2. The numerical subscripts of the parameters denote the particular variables involved, and the alphabetic subscripts denote the categories of those variables in the same order.

11.6.3 AN ILLUSTRATION

Let us compute the main effects parameters of the model,

$\ln = e_{ij} = u + u_1(i) + u_2(j)$ for the religion and type of service data. The $\ln e_{ij}$ values are as shown in the following table.

Religion	In-patients	Out-patients	Total
Hindu	4.854	5.627	1.481
Muslim	3.037	3.810	6.847
Christian	2.180	2.952	5.132
Total	10.071	12.389	22.460

The estimate of the main effect parameter $u_1(i)$ is obtained by,

$u_1(i) = \frac{1}{2}(10.481) - \frac{1}{6}(4.854 + - - - - - + 2.952) = 5.241 - 3.744 = 1.497$

The first set of estimate main effects are as shown in the following table

	Religion	Type of Service
	$u_1(1) = 1.497$	$u_2(1) = -0.386$
Category	$u_1(2) = -0.320$	$u_2(2) = 0.386$
	$u_1(3) = -1.177$	

The sum of the estimates for each variable is zero. The size of the effects simply reflects the size of the marginal totals. That is of the parameters $u_1(i)$, $u_1(1)$ is the largest since the first category (Hindu) of variable 1 (Religion) has the largest marginal total among those of the variable. Similarly of the parameters $u_2(j)$, $u_2(2)$ is the largest.

11.6.4 VARIABLES ARE NOT INDEPENDENT

When the two variables are not independent, the log-linear model is given by,

$$\ln e_{ij} = u + u_1(i) + u_2(j) + u_{12}(ij)$$

where $u_{12}(ij)$ represents the interaction effect between level i and j of variables 1 and 2, respectively. Interaction effects are measured as deviations and we have,

$$\sum_{j=1}^{c} u_{12}\left(ij\right) = 0 \text{ and } \sum_{i=1}^{r} u_{12}\left(ij\right)$$

Estimation of the interaction effects would be useful in identifying those categories responsible for any departure from independence.

KEYWORDS

- **Contingency table**
- **Non-parametric tests**
- **Test statistics**

CHAPTER 12

CORRELATION ANALYSIS AND REGRESSION ANALYSIS

CONTENTS

12.1   TWO-QUANTITATIVE VARIABLES CORRELATION COEFFICIENTS

12.1.1   SCATTER DIAGRAM

The primary step in correlation analysis is to present the bivariate data (two-variable data) on a graph sheet and grasp the type of relationship. Such a graphical representation is known as scatter diagram or dot diagram. The scatter diagrams showing different types of relationship are given in Figure 12.1.

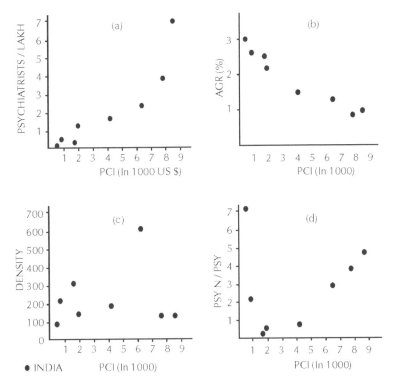

FIGURE 12.1   The dot diagram (a) shows the positive linear relationship between per capita income (PCI) and the number of psychiatrists per one lakh population of eight countries. The dot diagram (b) shows the negative linear relationship between PCI and annual growth rate (AGR) of population. The dot diagram (c) shows no relationship between PCI and the density of population. The dot diagram (d) shows curvilinear relationship between PCI and number of psychiatric nurses per psychiatrist manpower.

12.1.2 PEARSON CORRELATION COEFFICIENT

We limit our study to linear relationship and the theory built upon this assumption is known as linear or simple correlation. The product movement correlation coefficient or Pearson correlation coefficient measures such relationship. It is given by

$$r = \frac{\sum xy - n\bar{x}\,\bar{y}}{\sqrt{\left(\sum x^2 - n\bar{x}^2\right)\left(\sum y^2 - n\bar{y}^2\right)}}$$

The r lies between -1 and 1.

$r = 1$ means perfect positive linear relationship

$r = -1$ means perfect negative linear relationship

$r = 0$ means no linear relationship

The standard error of r is given by

$$SE\,(r) = \sqrt{\frac{1 - r^2}{(n - 2)}}$$

12.1.2.1 TEST OF SIGNIFICANCE

The procedures for testing a population correlation coefficient (ρ) equals zero is given below:

H_0: $\rho = 0$ against H_1: $\rho \neq 0$

The test statistic is given by

$$t = \frac{r\sqrt{n - 2}}{\sqrt{1 - r^2}} \quad \text{with df} = (n - 2)$$

Example: Let us suppose that the marks scored in mathematics (x), and the marks scored in statistics (y) by a group of five students are given in the following table.

Student	Maths (x)	Statistics (y)
1	4	7
2	2	4
3	3	1
4	0	3
5	6	5

For this data, the computation table is given below:

Student	x	y	x^2	y^2	xy
1	4	7	16	49	28
2	2	4	4	16	8
3	3	1	9	1	3
4	0	3	0	9	0
5	6	5	36	25	30
Total	15	20	65	100	69

Now,

$$r = \frac{69 - 5 \times 3 \times 4}{\sqrt{(65 - 5 \times 3^2)(100 - 5 \times 4^2)}} = \frac{69 - 60}{\sqrt{20 \times 20}} = \frac{9}{20} = 0.45$$

12.1.2.2 TEST OF SIGNIFICANCE

$$t = \frac{0.45\sqrt{5-2}}{\sqrt{1-0.45^2}} = \frac{0.45\sqrt{3}}{\sqrt{1-0.2025}} = 0.872, \text{ with df} = (5-2) = 3$$

Since the calculated value is less than the table value of $t(3) = 3.18$, the null hypothesis is accepted ($P > 0.05$, not significant).

12.2 RANK CORRELATION COEFFICIENTS

12.2.1 SPEARMAN RANK CORRELATION COEFFICIENT

The Spearman rank correlation coefficient is derived for the situation in which the observed values of both the variables are expressed as ranks. It is given by

$$r_s = 1 - \frac{6 \sum d_i^2}{n(n^2 - 1)}$$

The d_i is the difference between the ranks of the two variables of the ith individual. The r_s lies between -1 and 1.

Example: Let us calculate the Spearman rank correlation coefficient for the data used to demonstrate the Pearson correlation coefficient. The computation table is as given below:

Student	x	y	Ranks of		d_i	d_i^2
			x	y		
1	4	7	4	5	-1	1
2	2	4	2	3	-1	1
3	3	1	3	1	2	4
4	0	3	1	2	-1	1
5	6	5	5	4	1	1
Total						8

Now,

$$r_s = 1 - \frac{6 \times 8}{5(5^2 - 1)} = 1 - \frac{6 \times 8}{5 \times 24} = 0.60$$

12.3 TWO-QUALITATIVE VARIABLES CORRELATION COEFFICIENTS

12.3.1 CONTINGENCY COEFFICIENT

The first step in calculating the contingency coefficient is to present the data in a contingency table. The contingency coefficient is computed for the data in which both the variables are measured in nominal level. It is given by

$$CC = \sqrt{\frac{\chi^2}{\chi^2 + n}}$$

where χ^2 is the chi-square value. The CC lies between 0 and less than 1.

Example: The 2 × 2 contingency table for calculating the CC for sex and marital status of 40 patients (data in Appendix I) is given below:

Sex	Single	Married	Total
Male	14	10	24
Female	7	9	16
Total	21	19	40

For this data,

$$CC = \sqrt{\frac{0.819}{0.819 + 40}} = 0.142$$

12.3.2 PHI COEFFICIENT

The phi-correlation coefficient is computed for the data in which both the variables being correlated are genuine dichotomous, and the classes are separated by a real gap between them. In order to compute phi-coefficient, the 2 × 2 contingency table has to be prepared. The phi-coefficient is given by (with usual notations),

$$\varnothing = \frac{(ad - bc)}{\sqrt{(a+b)(c+d)(a+c)(b+d)}}$$

where a, b, c, and d are the frequencies in the 2 × 2 contingency table. The $\varnothing$ lies between −1 and 1.

Example: Association between sex and marital status of the 40 patients used to demonstrate the contingency coefficient is calculated as follows:

$$\varnothing = \frac{(14 \times 9 - 10 \times 7)}{\sqrt{24 \times 16 \times 21 \times 19}} = 0.142$$

12.3.3 BISERIAL CORRELATION COEFFICIENT

The biserial correlation coefficient is developed for the situation in which both the variables are measured in interval scales or continuously measured, but one of them is, for some reason, reduced to two categories. The point biserial correlation coefficient (r_{pb}) is appropriate when one of the two variables is a genuine dichotomy. A special formula is provided which does not resemble the basic pearson formula. It reads,

$$r_{pb} = \frac{\left(M_p - M_q\right)\sqrt{pq}}{\sigma_x}$$

where M_p is the mean of the continuous variable for the higher group in the dichotomous variable, M_q is the mean of the continuous variable for the lower group in the dichotomous variable, p is the proportion of the number of cases in the higher group of the dichotomous variable, q = (1-p), and σ_x is the standard deviation of the continuously measured variable for the whole sample. The reading ability test scores (x) and the gender of the five students are presented below.

Students	x	Gender
A	4	male
B	2	male
C	3	female
D	0	female
E	6	male

In this bivariate data, $M_p = 4$, $M_q = 1.5$, p = 0.6, q = 0.4 and $\sigma_x = 2$. Then

$$r_{pb} = \frac{\left(4 - 1.5\right)\sqrt{0.6 \times 0.4}}{2} = 0.612$$

The male gender is associated with higher scores, and the association is 0.612.

12.4   MEASURES OF RELATIVE RISKS

In observational longitudinal studies such as retrospective studies and prospective studies, we attempt to measure the relative risk of various risk factors on several diseases/disorders. For analysis of such data, the 2 × 2 risk and disease table has to be prepared as shown below.

		Disease		Total
		Present	Absent	
Risk	Present	a	b	(a + b)
	Absent	c	d	(c + d)

12.4.1 RISK DIFFERENCE

The risk difference is defined as follows:

RD = (risk in the exposed group-risk in the normal group)

$$= \left(\frac{a}{a+b}\right) - \left(\frac{c}{c+d}\right)$$

The RD lies between -1 and 1.

12.4.2 RISK RATIO

The risk ratio is defined as follows:

$$RR = \frac{\text{Incidence of the disease among exposed}}{\text{Incidence of the disease among non-exposed}}$$

$$= \left(\frac{a}{a+b}\right) / \left(\frac{c}{c+d}\right)$$

The RR lies between zero and infinity.

12.4.3 ODDS RATIO

The odds ratio is defined as follows:

$$OR = \frac{\text{Odds that exposed individuals will have disease}}{\text{Odds that non-exposed individuals will have disease}}$$

$$= \left(\frac{a}{b}\right) / \left(\frac{c}{d}\right) = \frac{ad}{bc}$$

The OR lies between zero and infinity.

OR < 1 means the exposed is protected
OR = 1 means there is no effect
OR > 1 means the exposed is harmful
Thus, OR indicates the number of times the risk is harmful in comparison with the normal.

12.4.4   TEST OF SIGNIFICANCE OF ODDS RATIO

The procedure for test of significance of OR equals 1 is given by,
H_0: OR = 1 against H_1: OR $\neq$ 1
We have the transformation, $\lambda = l_n$ (OR)
Now, H_0: $\lambda = 0$ against H_1: $\lambda \neq 0$
The test statistic is given by

$$Z = \frac{\hat{\lambda}}{\sqrt{var(\hat{\lambda})}}$$

where $\hat{\lambda}$ is based on the sample values.

$$Var(\hat{\lambda}) = \left(\frac{1}{a} + \frac{1}{b} + \frac{1}{c} + \frac{1}{d} \right)$$

Example: Let us suppose that the data are given in the following 2 × 2 risk and disease table.

		Disease		Total
		Present	Absent	
Risk	Present	13	9	22
	Absent	1	13	14

For this data,
$$RD = \frac{13}{22} - \frac{1}{14} = 0.591 - 0.071 = 0.520$$

$$RR = \frac{13}{22} \Big/ \frac{1}{14} = 0.591/0.071 = 8.27$$

$$OR = \frac{13}{9} \Big/ \frac{1}{13} = (13 \times 13)/9 = 169/9 = 18.78$$

The test of significance of OR is carried out as follows:

$$\lambda = l_n (OR) = l_n (18.78) = 2.933$$

$$\text{Var}(\hat{\lambda}) = (\tfrac{1}{13} + \tfrac{1}{9} + \tfrac{1}{1} + \tfrac{1}{13}) = 0.0769 + 0.1111 + 1.0000 + 0.0769 = 1.265$$

$$Z = \frac{2.933}{\sqrt{1.265}} = 2.61$$

Since the calculated value is more than 2.58, we reject the null hypotheses ($P < 0.01$, highly significant).

12.5   REGRESSION ANALYSIS

The main purpose of regression analysis is to predict the value of the dependent variable (y) when the value of the independent variable (x) is given. The analysis describes the dependency of a variable on independent variables, suggests possible cause-and-effect relationship between factors, and explains some of the variation of the dependent variable by the independent variables by using the latter as a control. For example, the studies such as (1) dose of amitriptyline on the reduction of depression and (2) effect of urbanization on the increase of prevalence of neurotic disorder of a defined population.

12.5.1   *LINEAR REGRESSION ANALYSIS*

In linear regression analysis (one-independent variable), a straight line passing through the data in dot diagram is to be fitted.

12.5.1.1   *FITTING LINEAR REGRESSION EQUATION*

The straight line equation is given by

$$y = a + bx$$

where a is the intercept

b is the slope of the straight line called regression coefficient

The parameters a and b are determined in such a way that the sum of squares of deviations of the observed values from their regression estimates is minimum. This is a typical problem in differential calculus. The values for a and b are given by

$$a = \bar{y} - b\,\bar{x}$$

$$b = \frac{\sum xy - n\bar{x}\bar{y}}{\sum x^2 - n\bar{x}^2}$$

Hence,

$$\hat{y} = a + bx$$

where $\hat{y}$ is the predicted value of the dependent variable for the given value of the independent variable x.

Example: Let us suppose that the dosage (x) and the improvement score (y) of a group of 5 students are given below:

Patient	Dosage	Improvement
1	4	7
2	2	4
3	3	1
4	0	3
5	6	5

For this data,

$$b = \frac{69 - 5 \times 3 \times 4}{65 - 5 \times 3^2} = \frac{9}{20} = 0.45$$

$$a = 4 - 0.45 \times 3 = 4 - 1.35 = 2.65$$

Hence, the regression equation is given by

$$\hat{y} = 2.65 + 0.45x$$

That is, the predicted improvement score $= 2.65 + 0.45 \times$ dose

Patient	Dosage (x)	Improvement (y)	Improvement ($\hat{y}$) Estimate	Residuals $(y-\hat{y})$
1	4	7	4.45	2.55
2	2	4	3.55	0.45
3	3	1	4.00	−3.00
4	0	3	2.65	0.35
5	6	5	5.35	−0.35
Total	15	20	20.00	0

The linear regression line is given in Figure 12.2.

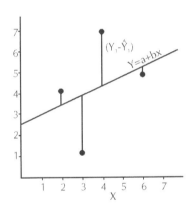

FIGURE 12.2   Dot diagram showing regression line of y on x.

Let us suppose that a patient is given a dose 1. Then, his predicted improvement score is determined as follows:

$\hat{y} = 2.65 + 0.45 \times 1 = 2.65 + 0.45 = 3.10$, approximately equal to 3

12.5.2   LOGISTIC REGRESSION ANALYSIS

The logistic regression is used to predict a discrete outcome. If the dependent variable has only two outcomes, the method is called binary logistic regression. If there are more than two outcomes associated with the dependent variable, the method is called multinomial logistic regression. The logistic regression computes the log (natural) odds for a particular outcome. The odds of an outcome are given by the ratio of the probability of it happening and not happening. That is,

$$\text{Odds} = \frac{P}{1-P}$$

where P is the probability of an event. We calculate

$$l_n\left(\frac{P}{1-P}\right)$$

The values range from $-\infty$ to ∞. A positive value indicates that odds are in favor of the event and the event is likely to occur, whereas a negative value indicates that odds are against the event and the event is not likely to occur. For establishing model fit, SPSS calculates chi-square values based on log-likelihood values.

KEYWORDS

- **Chi-square value**
- **Contingency table**
- **Correlation coefficient**
- **Regression analysis**

CHAPTER 13

RELIABILITY ANALYSIS AND VALIDITY ANALYSIS

CONTENTS

13.1   SCALING TECHNIQUES

The simplest way of measuring one's attitude is to ask him to rate his strength by himself. This can be done by constructing a scale by using summated scaling method or equal interval scaling method.

13.1.1   SUMMATED SCALING TECHNIQUE

In this method of measuring one's attitude is by presenting him with a number of attitude statements of varying intensity. The steps involved in construction and administration of summated scale are as follows:

Step 1: About 50 items are selected from the bank of items measuring the attitude/trait. These items must be valid and reliable.

Step 2: About 50 percent of the items are positively worded and the remaining 50 percent of the items are negatively worded.

Step 3: About 3–7 response categories ranging from low to high intensity of each item are to be adapted.

Step 4: The positively worded items must be measured in one order and then negatively worded items must be measured in the reverse order.

Step 5: The respondents are asked to tick the response categories for each item in the scale.

Step 6: We sum the scores of these items to portray their position of intensity.

Step 7: Transform the scores into ordinal level to facilitate appropriate analysis.

13.1.2   EQUAL-INTERVAL SCALING TECHNIQUE

In equal-interval scaling (Thurstone scaling), the respondents are presented with a hypothetical range of attitudes from extreme favorableness to extreme unfavorableness graphically or pictorially. The steps involved in construction and administration of Thurstone scale are as follows:

Step 1: About 50 items are selected from the bank of items.

Step 2: About 20 judges are asked to sort out these items into about 7 categories ranging from high to low intensity.

Step 3: The weight of each item is based on the average of the categories into which it has been sorted out by the judges.

Step 4: About 20 items are selected for use in the final questionnaire.

Step 5: The respondents are asked to select two or three items that best reflect their sentiments toward the psychological objects in question.

Step 6: We calculate the mean of these items to portray their position along the continuum of intensity.

Step 7: The Thurstone scaling measures lend themselves to interval level of measurement.

13.2 RELIABILITY ANALYSIS

Reliability is the consistency of a set of measurements or of a measuring instrument, often used to describe a test. There are several general classes of reliability estimates: test–retest reliability that includes intrarater reliability, internal consistency reliability, inter-rater reliability, and inter-method reliability.

13.2.1 TEST–RETEST RELIABILITY OF SCALES

It is the variation in measurements taken by a single person or instrument on the same items and under the same conditions. Thus, the test–retest method consists in submitting a group of individuals to a particular test and compiling their respective scores. After some time, the same test is repeated on the same group of individuals, and their scores are noted down. Then, the correlation between the two sets of related scores is obtained to measure the reliability of the test.

If the test is repeated immediately after the first, then the scores are likely to be improved on account of the memory effect, practice, and confidence. If sufficient time is given, then some other factors may come in such as growth in case of children, which may increase the scores. This method is generally used to measure the reliability of speed tests where speed is an important criterion. It is appropriate when the test is heterogeneous in the sense that different parts measure different traits (multi-dimensional scaling).

13.2.2   UNI-DIMENSIONAL SCALING

When items are used to form a scale, they need to have interval consistency. The items should all measure the same thing; therefore, they should be correlated with one another.

13.2.2.1   CRONBACH'S ALPHA

A useful coefficient for assessing interval consistency is Cronbach's alpha. It is given by

$$\text{Cronbach's alpha} = \frac{k\,\bar{c}}{(k-1)\bar{c} + \bar{v}}$$

where $\bar{c}$ is the mean of covariances
$\bar{v}$ is the mean of variances
k is the number of items

Example: Let us suppose that the rating scores on services rendered by a mental health institute (hospital) as rated by 5 raters on 3 items are given below:

Rater	Item 1	Item 2	Item 3
1	4	7	10
2	2	4	7
3	3	1	8
4	0	3	6
5	6	5	4

For this data, the variance–covariance matrix is given below:

Items	Item 1	Item 2	Item 3
Item 1	4	1.8	−0.6
Item 2	1.8	4	0.8
Item 3	−0.6	0.8	4

$$\text{Cronbach's alpha} = \frac{3 \times 0.667}{(2 \times 0.667) + 4} = 0.375$$

13.2.3   INTERNAL CONSISTENCY RELIABILITY OF SCALES

It assesses the consistency of results across items included in a test. The internal consistency reliability is usually measured by split-half method or the method of rational equivalence. These methods are appropriate for power test in which all examiners have time to furnish.

13.2.3.1   SPLIT-HALF METHOD

The split-half method consists in breaking the original test into two equivalent halves and computing the correlation coefficient (r_{hh}) between the scores in half tests. Then, the reliability coefficient for the whole test is determined in terms of self-correlation of the half tests by using Spearman–Brown prophecy formula,

$$r_{tt} = \frac{2r_{hh}}{1+r_{hh}}$$

In case of power test, the test items are arranged in increasing order of difficulty and then splitting them into two equivalent halves with odd and even numbered items provide a unique estimate of reliability.

13.2.4   INTER-RATER RELIABILITY OF CONCEPTS

Even if a concept such as schizophrenia is validated by certain genetic, biochemical, psychological, course or treatment variables, the inter-rater reliability assessment of such a concept is required to assume communicative value or to provide scientific evaluation.

13.2.4.1   TWO-RATERS AND DICHOTOMOUS RATINGS

The first step in the analysis of two-raters and dichotomous ratings is to present the data in a 2 × 2 dichotomous ratings table as shown below:

Rater A	Rater B	
	Agreed	Not-agreed
Agreed	a	b
Not-agreed	c	d

13.2.4.2 KAPPA COEFFICIENT

The inter-rater reliability is measured by using kappa coefficient. It is given by

$$K = \frac{(p_0 - p_c)}{(1 - p_c)}$$

where p_0 is the proportion of the sum of the main diagonal entries
p_c is the proportion of the sum of the chance number of agreements
Hence,

$$P_0 = \frac{(a + d)}{(a + b + c + d)}$$

$$p_c = \frac{E(a) + E(d)}{(a + b + c + d)}$$

The kappa coefficient lies between -1 and 1.
$k = 1$ means perfect agreement
$k = 0$ means no agreement
$k = -1$, means perfect disagreement.
Example: Let us suppose that the following data are the result of making diagnosis (schizophrenia or not) by two-raters A and B:

Rater A	Rater B		Total
	Schizophrenia	Non-schizophrenia	
Schizophrenia	40	20	60
Non-schizophrenia	10	30	40
Total	50	50	100

For this data,

$$p_0 = \frac{(40 + 30)}{100} = 0.70$$

The chance frequency for schizophrenia by both the raters is given by

$$E \text{ (schizophrenia and schizophrenia)} = \frac{(50 \times 60)}{100} = 30$$

The chance frequency for non-schizophrenia by both the raters is given by

$$E \text{ (non-schizophrenia and non-schizophrenia)} = \frac{(50 \times 40)}{100} = 20$$

Hence,

$$p_c = \frac{(30 + 20)}{100} = 0.50$$

$$k = \frac{(0.70 - 0.50)}{(1 - 0.50)} = \frac{0.2}{0.5} = 0.40$$

13.2.4.3   TWO-RATERS AND POLYCHOTOMOUS RATINGS

The two-raters and polychotomous ratings can be dealt with by applying the kappa coefficient (k) as described above.

13.3   VALIDITY ANALYSIS

The validity of a test is the accuracy with which it measures what it is supposed or intended to measure. The validity of a test is determined experimentally by obtaining the correlation coefficient between the test (x) and on some other independent standard test scores (y) called criterion. A criterion may be an objective measure.

13.3.1   VALIDITY CRITERIA

It is difficult to measure validity in practice because of several concepts involved in it.

13.3.1.1 FACE VALIDITY AND CONTENT VALIDITY

The face validity establishes whether the measuring device looks like it is measuring the correct characteristics. The face validity is done by showing the instrument to experts. The content validity refers to the extent to which a measurement reflects the specific intended domain of content. To establish content validity, researchers should first define the entire domain of their study and then assess if the instrument they are using truly represents this domain.

13.3.1.2 PREDICTIVE VALIDITY AND CONCURRENT VALIDITY

The predictive validity means that the measurement should be able to predict other measures of the same thing. For example, if a student is doing well on mathematics examination, he should also do well during his statistics examination. The predictive validity can also be a forecasting validity. The concurrent validity describes the present one.

13.3.1.3 CONSTRUCT VALIDITY AND FACTOR VALIDITY

The construct validity tries to establish an agreement between the measuring instrument and theoretical concepts. To establish construct validity, one must first establish a theoretical relationship and examine the empirical relationship. The factorial validity of a given test is defined by its factor loadings, and these are given by the correlation of the tests with each factor.

13.3.2 VALIDITY INDICES OF SCALE ITEMS

The selected items for a rating scale must be moderately difficult and have good discriminating indices.

13.3.2.1 DIFFICULT INDEX

The difficult index of an item in a scale indicates the proportion of candidates wrongly answered the item. Thus, the difficult index of the ith item is given by

$$Dif(i) = \frac{\text{Number of incorrect answers to the } i^{th} \text{ item}}{\text{Total number of responses to the } i^{th} \text{ item}}$$

The difficult index lies between 0 and 1. All the items in the scale must have moderate difficult indices. They are constructed in such a way that 50 percent of the candidates are expected to answer correctly, since the variation (information) is maximum at $p = 0.50$, which is $pq = 0.25$. That is, as p approximates to 0 or 1, the variance decreases toward the vanishing point.

13.3.2.2   DISCRIMINATION INDEX

The discrimination index of a test item is the power of the item to discriminate between people of favorable attitude and people of unfavorable attitude. The discrimination index of the ith item is given by

$$Dis(i) = \frac{\begin{array}{c}\text{Number of correct answers} \\ \text{to the } i^{th} \text{ item in the} \\ \text{favorable group}\end{array} - \begin{array}{c}\text{Number of correct answers} \\ \text{to the } i^{th} \text{ item in the} \\ \text{unfavorable group}\end{array}}{\text{Number of cases in the favorable group}}$$

where the favorable group consists of the first 27 percent and the unfavorable group consists of the last 27 percent of the individual arranged in decreasing order of aggregate of marks scored by them. The discrimination index lies between −1 and 1. The Dis (i) is 1 means that a person who passes the ith item would be in the favorable group and a person who fails that item would be in the unfavorable group. The discrimination index is said to be satisfactory if it is more than 0.30.

Example: The marks scored on each of the 5 items measuring the knowledge in statistics by 11 students are tabulated as given below:

Student	Item 1	Item 2	Item 3	Item 4	Item 5	Total
1	1	1	1	1	1	5
2	1	1	1	1	1	5
3	1	1	1	0	1	4
4	1	1	1	0	1	4
5	1	0	1	0	1	3
6	1	0	0	1	1	3
7	1	0	1	0	1	3
8	0	0	1	0	1	2
9	0	0	1	0	1	2
10	0	0	1	0	1	2
11	1	0	0	0	1	2
Total	8	4	9	3	11	35
Dif(i)	.27	0.64	.18	.73	0	
Dis(i)	.67	1.00	.33	.67	0	

Since the difficult index of the 5th item is 0 (far away from 0.50), and its discriminate index is also zero, it is not worth retaining it in the scale.

KEYWORDS

- **Inter-rater reliability**
- **Test–retest method**

CHAPTER 14

SURVIVAL ANALYSIS AND TIME SERIES ANALYSIS

CONTENTS

14.1 SURVIVAL ANALYSIS

The life table techniques are mainly applied in survival analysis, which deal with the expectation of occurrence of phenomenon according to time or the probability of occurrence of the events by time. The life table techniques were originally developed in the field of demography to express the duration of life experienced by a particular group of population during a particular period. The life table techniques may be applied to study the duration of stay of admitted patients in a mental institute (hospital stay tables). They may be applied to follow-up data of specific psychiatric patients maintained with certain specific drug/therapy (clinical life table). In hospital stay table, the admissions, patients remaining in the hospital, and the discharged cases are analogous to the births, living population, and deaths in the demography life table.

14.1.1 COHORT HOSPITAL STAY TABLE TECHNIQUE

The cohort hospital stay table is based on cohort life table techniques. Let us suppose that a group of patients who are admitted in a psychiatric emergency ward on a particular day were followed up in time and noted down how many of them remained at the end of the first day, how many of them remained at the end of the second day, and so on. From this basic data, the other components of cohort hospital stay table are computed as described below:

x: x is the stay in days. It is taken as 0, 1, 2, …. for a complete hospital stay table, and the values of x may be taken like 0, 5, 10, …. and so on for an abridged life table.

l_x: l_x is the basic component. It is the number of patients who stayed up to the end of the xth day.

d_x The d_x is the number of patients who got discharged between the xth day and the $(x+1)$th day. Thus,

$$d_x = l_x - l_{x+1}$$

L_x: L_x is the number of days stayed by the cohort between the xth day and the $(x+1)$th day. Hence, it is estimated by

$$L_x = l_x - \frac{1}{2} d_x$$

T_x: T_x is the number of days stayed by the cohort beyond the xth day. Thus, it is given by

$$T_x = L_x + L_{x+1} + ...$$

e_x: e_x is the expectation of further duration of stay of a patient who has already stayed for x days. It is estimated by

$$e_x = \frac{T_x}{l_x}$$

Example: Let us suppose that a group of 20 patients who were admitted in an emergency ward on a particular day were followed up in time and noted down how many of them remained at the end of the first day, how many of them remained at the end of the second day, and so on. This data denoted by l_x are given in the following table:

x	0	1	2	3	4	5	6
l_x	20	19	18	11	6	3	1

For this data, the cohort hospital stay table is prepared as shown below:

Cohort hospital stay table

x (1)	l_x (2)	d_x (3)	L_x (4)	T_x (5)	e_x (6)
0	20	1	19.5	68.0	3.4
1	19	1	18.5	48.5	2.6
2	18	7	14.5	30.0	1.7
3	11	5	8.5	15.5	1.4
4	6	3	4.5	7.0	1.2
5	3	2	2.0	2.5	0.8
6	1	1	0.5	0.5	0.5

The examples for calculating various components are as given below:

$d_2 = 18 - 11 = 7$

$L_2 = 18 - 3.5 = 14.5$

$T_2 = 0.5 + 2.0 + 4.5 + 8.5 + 14.5 = 30.0$

$e_2 = 30/18 = 1.7$

14.1.2 *CURRENT HOSPITAL STAY TABLE TECHNIQUE*

The basic component for a current hospital stay table is the duration of stay (d_x) of a group of patients discharged from the hospital. From this basic data, the other components of current hospital stay table are computed as described in the following table.

x: x is the interval start time in days.

d_x**:** d_x is the basic component. This is the number of patients discharged in the interval xth and $(x + 1)$th day.

l_x**:** This is the number of patients entering into the interval xth and $(x + 1)$th day.

$$l_x = l_{(x-1)} - d_{(x-1)}$$

p_x**:** This is the proportion of patients discharged between the xth and $(x + 1)$th day.

$$p_x = \frac{d_x}{l_x}$$

q_x**:** This is the proportion of patients staying in the interval xth and $(x + 1)$th day.

$$q_x = 1 - p_x$$

cd_x**:** This is the cumulative number of patients discharged up to xth day.

$$cd_x = \sum d_x$$

cp_x**:** This is the cumulative proportion of patients discharged up to the xth day.

$$cp_x = \frac{l_0 - cd_x}{l_0}$$

pd$_x$: This is the probability density of discharge.

$$pd_x = cp_{(x-1)} - cp_x$$

HR$_x$: This is known as hazard ratio. It is given by

$$HR_x = \frac{d_x}{(l_x - 0.5d_x)}$$

Example: Let us suppose that the number of days stayed by 20 patients is given below:

Duration of Stay (d_x)	0	1	2	3	4	5	6
Frequency (patients)	1	1	7	5	3	2	1

The current hospital stay table is prepared and presented as shown below:

Current hospital stay table

x	d_x	l_x	p_x	q_x	cd_x	cp_x	pd_x	HR_x
(1)	(2)	(3)	(4)	(5)	(6)	(7)	(8)	(9)
0	1	20	.050	.950	1	.950	.050	.051
1	1	19	.053	.947	2	.900	.050	.054
2	7	18	.389	.611	9	.550	.350	.483
3	5	11	.455	.545	14	.300	.250	.588
4	3	6	.500	.500	17	.150	.150	.667
5	2	3	.667	.333	19	.050	.100	1.000
6	1	1	1.000	.000	20	.000	.050	2.000

Examples for calculating the various components are as given below:
$$l_2 = 20 - 2 = 18$$
$$p_2 = 7/18 = 0.389$$
$$q_2 = 1 - 0.3889 = 0.611$$
$$cd_2 = 1 + 1 + 7 = 9$$
$$cp_2 = (20 - 9)/20 = 0.550$$

$$pd_2 = 0.900 - 0.550 = 0.350$$

$$HR_2 = 7/(18 - 3.5) = 0.483$$

14.2 TIME SERIES ANALYSIS

In many instances, the events behave predictably over time. In such cases, we can use time series forecasting to predict what will happen next. That is, a time series analysis deals with the factors influencing variation of several phenomena according to time and thus aid in forecasting events. A time series is a set of statistical observations arranged in chronological order. That is, time series is a sample of repeated measures of a single variable observed at regular intervals over a period of time. The length of the intervals could be hours, daily, monthly, or yearly. What is most important is that it should be regular. We typically expect to find over or more of the common idealized pattern in a time series, often in combination with one another.

14.2.1 TIME SERIES COMPONENTS

14.2.1.1 TREND

The trend of the time series indicates whether the series increase or decrease. For example, the yearly number of registrations in a newly established mental health institute over a period of 20 years.

14.2.1.2 SEASONAL VARIATION

The seasonal variation represents that portion of a series which may be attributed to the time of the year. For example, (1) the monthly number of registrations in a mental health institute over a period of 20 years and (2) monthly incidence rates of affective disorders in a rural community over a period of 20 years.

14.2.1.3 CYCLICAL COMPONENT

The cyclical component of a time series analysis is a phenomenon that is observed more frequently than the duration of the time series. The phenomena may be attributed to the administrative reforms of the hospital, natural calamity in the surrounding community, etc. For example, the yearly suicides rates in India over a period of 20 years.

14.2.1.4 RANDOM FLUCTUATIONS

The random fluctuations or irregular fluctuations are the movements in the data that cannot be considered as one of the proceeding types.

14.2.2 DECOMPOSITION OF TIME SERIES COMPONENTS

It is rare to find a real time series which is a pure case of just one component. The time series components could be related in any manner, but the relationship is usually considered to be additive. That is, some observation Y is composed of the components given by

$$Y = T + S + C + R$$

14.2.2.1 TREND

The secular trend or linear trend may be determined by fitting a least-square straight line equation to the data. That is,

$$Y = a + bx$$

where x represents the time variable whose values are given serially from 1 to the number of observations.

14.2.2.2   SEASONAL VARIATION

In decomposing of seasonal variation, often the arithmetic means of the season are calculated from the residuals $(Y - T)$ to represent the seasonal variation.

14.2.2.3   CYCLICAL COMPONENT

The cyclical components can be determined by the method of moving averages to the residuals $(Y - T - S)$.

14.2.2.4   RANDOM FLUCTUATIONS

The random fluctuations are obtained by subtracting the trend, seasonal variation, and cyclical component from the data. That is, it is obtained by

$$R = (Y - T - S - C)$$

14.2.2.5   NUMERICAL DEMONSTRATION

The following table shows the bimonthly number of out-patients registrations of dissociation disorder patients over a period of 5 years at a general hospital psychiatric unit.

Year	January–February	March–April	May–June	July–August	September–October	November–December	Total
1996	5	7	3	5	3	6	29
1997	1	6	7	4	3	5	26
1998	2	4	6	2	4	2	20
1999	4	7	7	5	5	5	33
2000	2	9	6	10	7	2	36
Total	14	33	29	26	22	20	144

A scanning of the figures in the last column of the table concludes an upward trend of the series. The series gradually decrease up to the year

1998 and increase thereafter. The figures in the last row of the table indicate a seasonal peak during March–April. In determining the trend, x is taken from 1 (January–February 1996) to 30 (November–December 2000). It may be computed that

$$T = 3.9 + 0.06x$$

In decomposition of seasonal variation, the trend is subtracted from the observations (y) in order to get the sum of the seasonal variation, cyclical component, and random fluctuations as shown in the table. The resultants (sum of seasonal variation, cyclical variation, and random fluctuations) are as shown below:

Year	January–February	March–April	May–June	July–August	September–October	November–December
1996	1.0	2.9	−1.1	0.8	−1.2	1.7
1997	−3.3	1.6	2.6	−0.5	−1.6	0.4
1998	−2.7	−0.7	1.2	−2.8	−0.9	−3.0
1999	−1.0	2.0	1.9	−0.2	−0.2	−0.3
2000	−3.3	3.6	0.6	4.5	1.5	−3.1
Mean (seasonal)	−1.9	1.9	1.0	0.4	−0.5	−0.9

The last row of the table indicates the seasonal variation. The seasonal variation is subtracted from the data in the last table in order to obtain the sum of the cyclical variation and random fluctuations as shown below:

Year	January–February	March–April	May–June	July–August	September–October	November–December	Mean (cyclical)
1996	2.9	1.1	−2.2	0.5	−0.7	2.7	0
1997	−1.5	−0.3	1.5	−0.9	−1.1	1.3	−0.2
1998	−0.8	−2.6	0.2	−3.2	−0.4	−2.0	−1.5
1999	0.9	0.1	0.9	−0.5	0.3	0.7	0.4
2000	−1.5	1.7	−0.5	4.2	2.0	−2.7	0.6

The last column of the table indicates the cyclical component.

The graphical representation of the trend, seasonal variation, and the cyclical components are shown in the following Figure 14.1.

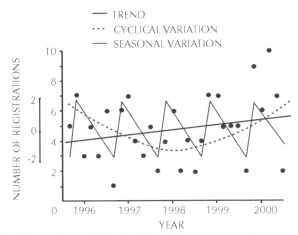

FIGURE 14.1   Time series analysis showing various components.

KEYWORDS

- **Cyclical component**
- **Seasonal variation**
- **Time series analysis**

CHAPTER 15

MULTIVARIATE STATISTICAL METHODS

CONTENTS

15.1 PROFILE TECHNIQUES

15.1.1 THREE-VARIABLES PROFILE

The profile techniques simultaneously plot data on several variables. Thus, a profile is best described as histogram on each variable, connecting between variables by identifying cases. The various objectives of profile techniques make it difficult to give complete explicit instructions for constructing profiles. It is especially useful in informally suggesting possible clusters of similar cases and also groups of similar variables.

Example: Let the percentage frequencies of male gender, percentage frequencies of high-income group, and percentage frequencies of admitted patients of five diagnostic categories of 409 psychiatric patients registered at NIMHANS hospital are as shown in the following table.

Variable	Organic Psychosis (14)	Substance Use Disorders (58)	Schizophrenia (124)	Affective Disorders (139)	Neurotic Disorders (74)
1. Male (%)	79	98	53	53	45
2. High income (%)	21	31	36	44	53
3. Rate of admission (%)	43	71	40	30	11

The initial analysis consists of single-variable summaries such as minimum and maximum percentages. The profile diagram is drawn as shown in Figure 15.1.

It can be grasped that the variable male gender and rate of admission are similar. Further, it may be grasped that schizophrenia and affective disorders are similar with respect to these three variables.

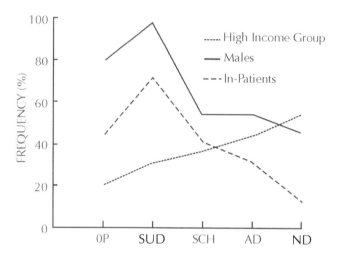

FIGURE 15.1   Profile showing percentage frequencies of diagnostic categories on three variables.

15.2   MULTIVARIATE CORRELATION COEFFICIENTS

15.2.1   PARTIAL CORRELATION COEFFICIENT

The correlation coefficient between two variables after eliminating the effect of the other variables is called partial correlation coefficient.

Example: The partial correlation coefficient between the fathers' heights and the sons' heights after eliminating the influence of mothers' heights. The partial correlation coefficient between the two variables (denoted by 1 and 2) after eliminating the influence of the third variable (denoted by 3) is given by

$$r_{12.3} = \frac{r_{12} - r_{13}r_{23}}{\sqrt{\left(1 - r_{13}^2\right)\left(1 - r_{23}^2\right)}}$$

The partial correlation coefficient lies between −1 and 1.

15.2.2 MULTIPLE CORRELATION COEFFICIENT

In multivariate data, the correlation between one variable and the combined influence of the other variables is known as multiple correlation coefficients. In trivariate data (three variables data), it is given by

$$R^2_{1(23)} = \frac{r_{12}^2 + r_{13}^2 - 2\,r_{12}r_{13}r_{23}}{(1 - r_{23}^2)}$$

The multiple correlation coefficient lies between 0 and 1.

Example: Let us assume the marks scored by 5 students in mathematics (denoted by variable 1), statistics (denoted by variable 2), and psychology (denoted by variable 3) as given below:

Student	Variable 1	Variable 2	Variable 3
1	4	7	10
2	2	4	7
3	3	1	8
4	0	3	6
5	6	5	4

For this data, the correlation coefficient matrix is computed and presented below:

	Variable 1	Variable 2	Variable 3
Variable 1	-	0.45	−0.15
Variable 2	-	-	0.20
Variable 3	-	-	-

For this data, the partial correlation coefficient between mathematics (variable 1) and statistics (variable 2) after eliminating the influence of psychology (variable 3) is calculated as follows:

$$r_{12.3} = \frac{0.45 - (-0.15)(0.20)}{\sqrt{(1 - (-0.15^2)(1 - 0.20^2))}} = \frac{0.45 + 0.03}{\sqrt{(1 - 0.0225)(1 - 0.04)}} = 0.4955$$

For this data, the multiple correlation coefficient between mathematics (variable 1) and the combined influence of statistics (variable 2) and psychology (variable 3) are calculated as follows:

$$R^2_{2(13)} = \frac{0.45^2 + 0.20^2 - 2 \times 045 \times 0.20(-0.15)}{1 - (-0.15)^2} = \frac{0.2025 + 0.04 + 0.027}{(1 - 0.0225)} = \frac{0.2695}{0.9775} = 0.2757$$

Hence,

$$R_{1(23)} = 0.525$$

15.3   MULTIPLE REGRESSION ANALYSIS

In multiple regression analysis, we have more than one independent variables $(x_1, x_2, \ldots x_k)$. The multiple regression analysis is based on the assumption that all the variables are normally distributed. A regression of k independent variables upon a dependent variable may be expressed as follows:

$$y = a + b_1 x_1 + b_2 x_2 + \ldots + b_k x_k$$

where a is a constant term

b_i are the simple partial regression coefficients

15.3.1   TWO-INDEPENDENT VARIABLES REGRESSION ANALYSIS

In case of two-independent variables $(x_1$ and $x_2)$, we have the regression equation as given below:

$$y = a + b_1 x_1 + b_2 x_2$$

where

$$a = \overline{y} - b_1 \overline{x}_1 - b_2 \overline{x}_2$$

$$b_1 = \frac{\left(\sum X_1 Y\right)\left(\sum X_2^2\right) - \left(\sum X_2 Y\right)\left(\sum X_1 X_2\right)}{\left(\sum X_1^2\right)\left(\sum X_2^2\right) - \left(\sum X_1 X_2\right)^2}$$

$$b_2 = \frac{\left(\sum X_2 Y\right)\left(\sum X_1^2\right) - \left(\sum X_1 Y\right)\left(\sum X_1 X_2\right)}{\left(\sum X_1^2\right)\left(\sum X_2^2\right) - \left(\sum X_1 X_2\right)^2}$$

$$X_1 = \left(x_1 - \bar{x}_1\right)$$

$$X_2 = \left(x_2 - \bar{x}_2\right)$$

$$Y = \left(y - \bar{y}\right)$$

Example: Let us assume that we have data of five students on two independent variables (IQ and personality scores) and dependent variable (success scores) as shown below:

Student	I Q (x_1)	Personality Score (x_2)	Success Score (y)
1	91	4	7
2	100	2	4
3	109	3	1
4	97	0	3
5	103	6	5
Mean	100	3	4

Computation table

Student	x_1	x_2	y	X_1	X_2	Y	X_1^2	X_2^2	$X_1 X_2$	$X_1 Y$	$X_2 Y$
1	91	4	7	−9	1	3	81	1	−9	−27	3
2	100	2	4	0	−1	0	0	1	0	0	0
3	109	3	1	9	0	−3	81	0	0	−27	0
4	97	0	3	−3	−3	−1	9	9	9	3	3
5	103	6	5	3	3	1	9	9	9	3	3
Total	500	15	20	0	0	0	180	20	9	−48	9

$$b_1 = \frac{(-48)(20) - (9 \times 9)}{(180 \times 20) - 9} = \frac{-10.41}{3591} = -0.290$$

$$b_2 = \frac{9 \times 180 - (-48)9}{(180 \times 20) - 9} = \frac{2052}{3591} = 0.5714$$

$a = 4 - (-0.29) \times 100 - (0.5714) \times 3 = 4 + 29 - 1.7142 = 31.2858$

$\hat{y} = 31.2858 - 0.29x_1 + 0.5714\,x_2$

That is, success score $= 31.2858 - 0.29$ IQ $+ 0.5714$ personality score

Now, the predicted values of the dependent variable ($\hat{y}$) are computed as shown in the following table:

Student	y	$\hat{y}$	$(y - \hat{y})$
1	7	7.25	−0.25
2	4	3.42	0.58
3	1	1.33	−0.33
4	3	3.15	−0.15
5	5	4.85	0.15
Total	20	20.00	0

The estimates are fairly close to the actual values and the correlation between the two is 0.996, which is the multiple correlation coefficient denoted by R. This is the maximum possible correlation between the dependent variable and any linear combination of the independent variables.

15.4   MULTIVARIATE ANALYSIS OF VARIANCE

15.4.1   TWO-GROUPS AND TWO VARIABLES

15.4.1.1   LAYOUT

The layout of multivariate analysis of variance (MANOVA) is as given in the example.

15.4.1.2   HYPOTHESES

The means of a sample on several variables may be thought of a single point in a space that has as many dimensions as there are variables. The MANOVA technique may be employed to test the significance of the difference between multivariate means of several groups.

15.4.1.3   TEST STATISTIC

The chi-square test statistic is given by

$$\chi^2 = -\left(\sum n_g\right) l_n \text{ with df} = (g-1)\, t$$

where g is the number of groups
   t is the number of variables (tests)

$$L = \frac{W}{T}$$

where W is the pooled within group sum of squares and sum of products matrix
   T is the total sum of squares and cross-products
   The L-criterion lies between 0 and 1. If $L = 0$, then there is no between groups variances or covariances. This means that every group has the same mean score on a particular variable and that this is true for all the variables.
   Example: Let us suppose that the marks scored by three arts graduates and the marks scored by three science graduates on two aptitude tests are given below:

Arts Graduate	Test 1	Test 2
Graduate 1	3	6
Graduate 2	4	4
Graduate 3	5	8
Mean	4	6

Science Graduate	Test 1	Test 2
Graduate 4	1	2
Graduate 5	2	0
Graduate 6	3	4
Mean	2	2

For this data, we calculated the sum of squares and sum of products matrix of each diagnostic groups.

$$W = W_1 + W_2$$

$$W_1 = \begin{bmatrix} 2 & 2 \\ 2 & 8 \end{bmatrix} \quad W_2 = \begin{bmatrix} 2 & 2 \\ 2 & 8 \end{bmatrix} \quad W = \begin{bmatrix} 4 & 4 \\ 4 & 16 \end{bmatrix}$$

$$T = T_1 + T_2$$

$$T_1 = \begin{bmatrix} 5 & 8 \\ 8 & 20 \end{bmatrix} \quad T_2 = \begin{bmatrix} 5 & 8 \\ 8 & 20 \end{bmatrix} \quad T = \begin{bmatrix} 10 & 16 \\ 16 & 40 \end{bmatrix}$$

$$L = \frac{W}{T} = \frac{(64-16)}{(400-256)} = \frac{48}{144} = \frac{1}{3}$$

Hence,

$\chi^2 = -6\, l_n(\tfrac{1}{3}) = -6\,(-1.0986) = 6.592$, with df $= (2-1)\,2 = 2$

Since the calculated value is more than the table value of $\chi^2(2) = 5.99$, we reject the null hypotheses ($P < 0.05$, significant). That is, we have the evidence that the two groups are drawn from populations that differ in their mean quantities of x or in their mean quantities of y or in some compound of the two.

15.5 ANALYSIS OF MULTI-DIMENSIONAL CONTINGENCY TABLES

All the tables that simultaneously present data of more than two qualitative variables are known as multi-dimensional contingency table. A three-dimensional $r \times c \times l$ contingency table has r rows, c columns, and l layer categories. The observed frequency in the ijk-th cell of the table is denoted by $O_{ijk}(i = 1, 2, ...r; j = 1, 2, ...c; k = 1, 2, ...l)$. By summing the O_{ijk} over different subscripts, various marginal totals may be obtained.

15.5.1 ANALYSIS OF THREE-DIMENSIONAL CONTINGENCY TABLE

2 × 2 × 2 Contingency Table

Variable j	Variable k				
	Category 1		Category 2		
	Category 1	Category 2	Category 1	Category 2	Total
Variable i Category 1	O_{111}	O_{121}	O_{112}	O_{122}	$O_{1..}$
Category 2	O_{211}	O_{221}	O_{212}	O_{222}	$O_{2..}$
Total	$O_{.11}$	$O_{.21}$	$O_{.12}$	$O_{.22}$	
Total	$O_{..1}$		$O_{..2}$		Total

15.5.1.1 HYPOTHESES

The null hypotheses state that the three variables are mutually independent. That is,

$$H_0: P_{ijk} = P_{i..} P_{.j.} P_{..k}$$

where P_{ijk} represents the probability of an observation occurring in the ijk-th cell, and $P_i P_j P_k$ are the marginal probabilities of the row, column, and layer variables, respectively. When H_0 is true, the expected values are obtained as follows:

$$e_{ilk} = \frac{Oi.. \ O.j. \ O..k}{n^2}$$

15.5.1.2 TEST STATISTIC

It is given by

$$\chi^2 = \sum\sum\sum \frac{(O_{ijk} - e_{ijk})^2}{e_{ijk}} \quad df = rcl - r - c - l + 2$$

Example: The following table shows data concerning classroom behavior of 97 school children. Three variables are involved in the table. The first variable is a teacher's rating of classroom behavior classified into non-deviant and deviant. The second variable is the home risk index based on several items such as overcrowding thought to be related to deviant behavior. The third variable is an index of the adversity of school conditions based on items such as pupil turnover.

	Adversity of School Conditions (k):				Total
	Low		High		
Home risk index (*j*):	NAR	AR	NAR	AR	Total
Classroom non-deviant Behavior (i):	16	7	20	37	80
Deviant	1	1	4	11	17
Total	17	8	24	48	97
Total	25		72		97

NAR, not at risk and AR, at risk.

We have computed the expected values as shown in the following example:

$$e_{111} = \frac{80 \times 41 \times 25}{97 \times 97} = 8.72$$

The full set of expected values is given below in the following table:

	Adversity of School Conditions (k):				Total
	Low		High		
Home risk index (j):	NAR	AR	NAR	AR	Total
Classroom non-deviant Behavior (i):	8.72	11.90	25.10	34.28	80
Deviant	1.85	2.53	5.33	7.28	17
Total	10.57	14.43	30.43	41.56	97

$\chi^2 = 12.905$ df $= 4$, $P < 0.05$, significant.

KEYWORDS

- **Correlation coefficient**
- **MANCOVA**

CHAPTER 16

CLUSTER ANALYSIS

CONTENTS

16.1 FEATURES OF CLUSTER ANALYSIS

16.1.1 FORMAL INTRODUCTION

Cluster analysis encompasses many diverse techniques for discovering structure within complex bodies of data. In a typical situation, one has a sample of data units (e.g., psychiatric patients) each described by scores on selected variables, such as achievement test scores. The objective is to group either the data units or the variables into clusters such that the elements within a cluster have a high degree of 'natural association' among themselves while the clusters are 'relatively distinct' from one another. The approach to the problem and the results achieved depend principally on how the investigator chooses to give operational meaning to the phrases 'natural association' and 'relatively distinct'.

16.1.1.1 CLASSIFICATION, IDENTIFICATION, AND DISCRIMINANT ANALYSIS

Classification, identification or assignment, and discriminant analysis are key concepts closely linked to cluster analysis.

The term classification may refer to an end product or to a process. As an end product, a classification is a systematic scheme or arrangement of classes already known or established. As a process, classification consists of sorting individual objects into well-known classes.

Identification is the allocation of individual objects to established classes on the basis of specific criteria. In psychiatry and medicine the same process is called diagnosis, and it refers to the identification of a familiar disorder from the symptoms presented by the patient.

16.1.2 FUNCTION OF CLUSTER ANALYSIS

For a variety of research goals, researchers need to find out which objects in a set are similar and which objects are dissimilar. The best known of these research goals is the making of classifications. One reason that cluster analysis is also useful is that researchers in all fields need to make and revise classifications continuously. For example, psychiatric researchers need to make classification of mental patients based on psychiatric test scores in order to improve their understanding of mental illness and plan for its treatment.

16.1.2.1   SHEDDING LIGHT ON PREVIOUSLY MADE HYPOTHESIS

Clustering techniques may be useful in shedding light on the previously made hypothesis. For example, there has long been controversy over the classification of depressed patients. The issues involved here have been reviewed on a number of occasions. Several attempts have been made to establish the validity of classifying such patients into reactive and endogenous groups. More recently the problem has been tackled with some success by cluster analysis techniques.

16.1.2.2   PREDICTION

The clustering techniques may be useful to produce groups that form the basis of a classification scheme useful in later studies for predictive purposes of some kind. For example, a cluster analysis applied to data consisting of a sample of psychiatric patients may produce groups of patients who react differently when treated with some drug. Thus, it enables the investigator to decide whether a drug is suitable for a particular type of patients. Such a procedure is used in an investigation of the usefulness of amitriptyline in the treatment of depression.

16.1.2.3   OTHER PURPOSES

In addition to these goals, the researchers may use cluster analysis (1) to summarize and display the data, to name the resultant groups and to give explanations, (2) to explore the data and to generate hypotheses, (3) to investigate the usefulness of conceptual scheme for grouping entities and (4) to fit models. The data is summarized by referring to the properties of the clusters rather than the properties of the individuals. The data is displayed so as to bring the suitable difference more apparent. All the items in the same cluster will be given the same name. If some items have some common properties in a cluster, the other items are expected to have the same properties.

16.1.3   HISTORICAL BACKGROUND

Classifying people into types was an ancient pastime. The Hindus used sex and physical and behavioral characteristics to classify people into six types that they have designated by the names of animals. Galen defined nine temperamental types that were assumed to relate to a person's susceptibility to various diseases and to individual differences in behavior. Most of the early works on classification were in the fields of biology and zoology, where it is more generally known as taxonomy.

The advent of high-speed electronic computers and the fundamental importance of classification as a scientific procedure accelerated the applications of cluster analysis in many different fields. During the last 25 years, statisticians have become aware of the need for a more formal approach to cluster analysis and attempts have been made to formulate precise statistical models and to adapt a more rigorous approach to classification problems. It is natural that the literature on cluster analysis should be found in a variety of journals, ranging from electrical engineering to biology and from library sciences to psychiatry. The social sciences literature on clustering reflects a bewildering and often contradictory array of terminology, methods, and preferred approaches.

16.1.4   STRUCTURE TYPE

Generally speaking, there are three clustering structures: the tree, the partition, and the clump. A cluster is a subset of a set of objects. A tree is a family of clusters that has the property that any two clusters are either disjointed or one includes the other.

A partition is a family of clusters that has the property that any item just belongs to one of the partitions. The partitioning model for cases by variable data is that within each cluster each variable is constant over cases. This can be made more probabilistic by allowing cases to deviate from their cluster means. The partitioning model for distance data may be that the distances of any item with its cluster is less than its distance with any other cluster. Thus, a partition with a set of all items added is a tree. A clump is a family of clusters that has the property that the classes can overlap.

The choice of the clustering structure depends on the purpose of classification, size, and nature of data and the mode of search adapted.

16.1.5 IMPORTANCE IN PSYCHIATRIC RESEARCH

The study of the diseases of the mind is more elusive than the study of the diseases of the body, and the classification of such diseases is in an uncertain state. There is a general agreement on the existence of some psychiatric diseases like schizophrenia, paranoid, and depression, but clear cut diagnostic criteria are not available. The characteristic difficulty of classification of mental illness is the subtle, subjective, and varying nature of the symptoms. Numerical techniques gained more importance here than mere medical diagnosis. The standard numerical techniques consist of obtaining the data on demographic, psycho-social and clinical variables. The original data may be quite voluminous, but they are usually reduced by averaging over groups of items or by factor analysis.

Some papers seek the clusters of patients. Other papers seek the cluster of symptoms, and a few seek the clusters of variables and cases simultaneously.

16.1.6 CLUSTER ANALYSIS AND FACTOR ANALYSIS

Prior to 1955, cluster analysis represented a method of grouping variables instead of objects into subjects in order to define dimensions or factors and developed several such techniques to avoid using factor analysis, a relatively complex statistical procedure that was laborious and time-consuming before the computer revolution. Tryon conceived of his cluster analysis as a poor man's factor analysis.

In factor analysis, the sample from populations studied are known to be multimodel. In factor analysis, however, the multivariate distributions are ordinarily unimodel. If they were multimodel, interpretation of factor would be different and conflicting. Again, in factor analysis the variance and covariance are partitioned among several extracted factors or sources. Factor analysis is called for when the goal is to simplify the attributes and express them in terms of a smaller number of common factors. In cluster analysis, on the other hand, the total variance of the entity is assigned to a discrete subgroups or clusters. Cluster analysis is most appropriate where discrete categories are sought. The cluster procedures reduce the complexity of a data set by sorting entities into a smaller number of homogenous subgroups.

16.1.7 CLUSTER ANALYSIS AND DISCRIMINANT FUNCTION ANALYSIS

Discriminant analysis is a process undertaken to differentiate between groups formed on a particular basis. The problem here is not to discover groups but to identify a set of characteristics that can significantly differentiate between groups. The process allows one to allocate new cases to one of the groups with the least amount of error. In contrast, the clustering process generates a new categorical scheme or recovers groups within a mixture of several populations.

16.1.8 LIMITATIONS

Clustering is vulnerable on two fronts: first, the experts in the applied field may think that detailed knowledge is more important than mere manipulations, and secondly, the techniques are not based on sound probability models and the results are poorly evaluated and unstable when evaluated.

Besides these limitations, some precautions have to be taken in the use of cluster analysis techniques. Different clustering techniques can and do generate different solutions to the same data set and hence a number of validation techniques have been developed to provide some relief for this problem.

Again, the strategy of cluster analysis is structure-seeking although its operation is structure-imposing and therefore the key to using cluster analysis is knowing when these groups are 'real' and not merely imposed on the data by the method.

16.2 MEASURES OF SIMILARITY

Majority of clustering techniques begin with the calculation of a matrix of similarity/distance between entities and therefore consideration is needed on the possible ways of defining these quantities. Indeed, many clustering techniques may be thought of as attempts to summarize the information on the relationship between entities so that these relationships can be easily comprehended and communicated obviously, the output of the clustering technique will only be meaningful as the input, viz: similarities/distances between the entities.

The quantitative estimation of similarity has been dominated by the concept of metrics. This approach to similarity represents cases as points in a coordinate space such that the observed similarities of the points correspond to metric distances between them. There are four standard criteria that can be used to judge whether a similarity measure is a true metric. They are symmetry, triangle inequality, distinguishability of non-identicals, and indistinguishability of identicals. Many, but not all, distance measures commonly used are metrics. A number of correlation measures are not metrics. Coefficients that are not metrics may not be jointly monotonic. Despite their obvious importance, metrics are not the only way to represent the similarity between objects. In social sciences, asymmetric similarity values are commonly used.

A similarity coefficient measures the relationship between the individuals, given the values of a set of P variates common to both. Similarly coefficients take values in the range zero to one.

There are two broad groups of measures, viz: association measures including correlation coefficient and distance measures. These measures are briefly described in this section and a small write up is given on the choice of similarity measures.

16.2.1 ASSOCIATION COEFFICIENTS

In many cases the variates are of the 'presence' and 'absence' type which may be arranged in the familiar two-way association table.

Individuals i	Individual j		
	Present	Absent	Total
Present	a	b	a+b
Absent	c	d	c+d
Total	a+c	b+d	N

Many different coefficients have been suggested for data of this type. Some of them are shown here.

1) Simple Matching Coefficient
   $(a+d) / (a+b+c+d)$
2) Jaccard's Coefficient
   $a / (a+b+c)$
3) $2a / (2a+b+c)$
4) $a / (a+b+c+d)$

Large number of association coefficients were proposed, mainly because of uncertainty over how to incorporate negative matches into the coefficients, and also whether or not the matched pairs of variables are equally weighted. The fact that the different coefficients take different values on the same set of data would be relatively unimportant if all the coefficients were jointly monotonic. Each set of data must be considered on its own merits, by the investigator who is more familiar with the material. The simple matching coefficients cannot be easily transformed into a metric. However, considerable effort has been devoted to the establishment of approximate confidence limits. If simple matching coefficient is used, some cases would appear very similar primarily because they both locked the same features rather than because of the features they did have were shared. In contrast, Jaccard's coefficient is concerned only with features that have positive co-occurrences. The association coefficients are used to measure the similarities between variables also.

16.2.2 CORRELATION COEFFICIENT

The commonly used measure of similarity between individuals is the Product Moment Correlation Coefficient. This measure is generally suggested when the similarity of profile shapes is considered more important than the similarity of average profile levels. This is because of the fact that the correlation is positive when two profiles are parallel irrespective of how far apart they are. Unfortunately, the converse is not true. Two profiles may not be parallel but they may have correlation of +1. All that is required for a perfect correlation is that one set must be linearly related to the second set. The correlation coefficient can be criticized on other grounds also. For example, in its computation between cases, the 'mean' over disparate variables of one object is used and this would seem to have little meaning.

The similarity between profiles can be decomposed into three parts, viz: the shape, scatter, and elevation. The shape is the pattern of dips and rises across the variables. The scatter is the dispersion of the scores around their averages, and elevation is the mean score of the case over all of the variables. The Pearson's r is sensitive to shape because of its implicit standardization of each case across all variables. This property is of special importance to disciplines such as psychology in which data is often described in terms of profiles. It is possible that misleading results can be obtained if the effects of dispersions and elevation on profile data are not considered. Finally the Pearson's coefficient often fails to satisfy the triangle inequality.

16.2.3   DISTANCE MEASURES

The distance measures are the measures of dissimilarities. Normally they have no upper bounds and they are scale dependent.

16.2.3.1   EUCLIDEAN DISTANCE

The Euclidean distance d_{ij} between two individuals i and j is given by,

$$d_{ij} = \sqrt{\sum \left(x_{ik} - x_{jk} \right)^2}$$

where x_{ik} is the value of the kth variable for the ith individual. The Euclidean metric (distance) is the commonly used measure of distance in cluster analysis. The Euclidean distance used on a raw data may be very unsatisfactory since it is badly affected by the change of scale of a variable. Because of this, variables are frequently standardized to have zero mean and unit variance. Although this has problems, the Euclidean distance calculated from the standardized variables will preserve relative distances.

16.2.3.2   ABSOLUTE DISTANCE

In the absolute distance or city block distance, the distance d_{ij} between individuals i and j is given by,

$$d_{ij} = \left| x_{ik} - x_{jk} \right|$$

16.2.4   CHOICE OF MEASURES

The question as to which measure of similarity/distance should be used is an important consideration since different measures may lead to different results. The correlation coefficient and the Euclidean distance are generally used measures of similarity and distance, respectively, but they are not necessarily the most suitable measures in all situations. The choice of the correct measure to be used would be much simpler if we had prior knowledge of the structure of data. Thus the difficulties associated with finding ideal similarity measure give some advantages to the clustering techniques which do not rely heavily on such choices. Skinner (1978) has proposed a useful and feasible way in which both the correlation and the Euclidean distance are used to calculate the similarity of profile data so that it is possible to determine which of the factors (shape, size, and scatter) have contributed to the estimation of similarity.

16.3   CHOICE OF VARIABLES

16.3.1   CONFIRMATORY WITH THEORY

The choice of a particular set of variables used to describe each entity reflects the investigator's judgment of relevance for the purpose of the classification. The question is whether the correct ones have been chosen in the sense that they are relevant to the type of classification being sought. It is important to bear in mind that the initial choice of the variables itself is a categorization of the data which has no statistical guidelines. However, ideally, variables should be chosen within the context of an explicitly stated theory that is used to support the classification.

16.3.2   TRANSFORMATION

A logarithmic or the transformation is often performed, when the normality assumption of a variable is violated. If the data are not of the same scale values, they are commonly standardized to a mean of zero and to unit variance. There is some controversy, however, as to whether standardization should be a routine procedure in cluster analysis. Standardization to unit variance and mean of zero can reduce the differences between groups on those variables that may well be the best discriminators of group differences. It would be far more appropriate to standardize variables within clusters, but this cannot be done until the cases have been placed into clusters.

Variables in multivariate data sets may have different distribution parameters across groups; thus standardization may not constitute an equivalent transformation of these variables and could possibly change the relationships between them. However, his Monte Carlo studies of the effects of standardization on subsequent analyses using the correlation coefficient and various hierarchical clustering methods did not reveal substantial difference between the use of standardized versus non-standardized variables in the resulting classifications. Standardization appears to have only a minor effect on the result of a cluster analysis. Others, have shown that standardization did not have a negative effect on the adequacy of the results of a cluster analysis when compared to an "optimal" classification of the cases under study.

The situation regarding standardization is far from clear. Users with substantially different units of measurement will undoubtedly want to standardize them, especially if a similarity measure such as Euclidean distance is to be used. The decision to standardize should be made on a problem-to-problem basis, and users should be aware that results may differ solely on the basis of this factor, although the magnitude of the effect will vary from data set to data set.

16.3.3   WEIGHTAGE TO VARIABLES

Allied to the measurement of similarities is the question of weighing the variables. This means giving different weights to the variables in the determination of the similarity coefficient. The reasoning that the weights can be based on intuitive judgments of what is important may merely reflect the existing classification of the data. In many instances, it may make sense to weigh a particular variable a priori if there is good theoretical reasons for doing this and there are well-defined procedures under which weighing can be made.

16.3.4 NUMBER OF VARIABLES

The number of variables used to measure each entity is generally large and the amount of computer time taken increases dramatically with an increase in the number of variables. Perhaps the simplest way to do this is to perform a principal component analysis on the data, and use the first few principal components scores as input variables to the clustering procedures. Such a procedure is most unlikely to lead to seriously misleading results.

Such clustering techniques assume that the variables are uncorrelated within clusters and hence may give 'spurious' solutions in cases where the assumption is untrue. The application of principal component analysis on raw data not only reduces the dimensionality of the data, but also creates new and uncorrelated variables.

16.4   HIERARCHICAL METHODS

In hierarchical methods, the classes themselves are classified into groups and the process is repeated at different levels to form a tree. There are two types of hierarchical methods, viz: the agglomerative techniques and the divisive techniques.

The agglomerative techniques proceed by a series of successive fusions of the n entities into groups, and the divisive techniques partition the set of entities successively into finer partitions. Thus, agglomerative techniques ultimately reduce the data to a single cluster containing all the entities, and the divisive techniques will finally split the entire set of data into groups each containing a single entity.

DENDROGRAMS

The results of both agglomerative and divisive techniques may be presented in the form of a dendrogram. A dendrogram is a two-dimensional diagram illustrating the fusions or partitions which have been made at each successive level.

The basic procedure with all hierarchical agglomerative methods is similar. They begin with the computation of a matrix of similarity/distances between the entities. At any particular stage, the technique fuses individuals or groups of individuals which are closest. Differences between techniques arise because of the different ways of defining similarity/distance between groups of individuals. Some of these techniques are suitable for use only when a distance matrix is used as the starting point. These methods search for $n \times n$ similarity matrix and subsequently merge the most similar cases. The distance measure between groups used in many agglomerative methods satisfy a recurrent formula for the distance between a group k, and a group (ij) formed by the fusion of two groups i and j. This formula is given by

$$d_k(i,j) = Ad_{ki} + Bd_{kj} + Cd_{ij} + D\left|d_{ki} - d_{kj}\right|$$

where d_{ij} is the distance between groups i and j and A, B, C, and D are the parameters whose values for a particular method are given.

16.4.1 SINGLE LINKAGE METHOD (SLINK)

In single linkage method, the groups initially consisting of single individuals are fused according to the distance between their nearest members, the groups with the smallest distance being fused first. Each fusion decreases the number of groups by one. For this method, the distance between groups is defined as the distance between their closest members. This method can be used with both similarity and distance measures. In terms of parameters for recurrent formula, SLINK specifies:

$$A = B = 0.50 \; ; C = 0 \; ; D = -0.50$$

The major drawback of this method is that it has the tendency to chain, or form long and elongated clusters. Here an entity is added to a cluster if it is highly similar to any member of that cluster. Because of this criterion, the resulting cluster can tend to resemble long chain when plotted in a multidimensional space. The chaining phenomenon had led most reviewers of SLINK to reject it as a preferable clustering procedure.

Some authors have pointed out that SLINK preserves the ultrametric inequality when representing similarity relationships. That is, given that i, j, and k refer to any three entities to be clustered, SLINK preserves the following relationship concerning the distance among three entities.

$$d_{(ij)} < \text{Max.} \, [d_{(ik)}/, d_{(jk)}]$$

Williams, et al (1971) have suggested that although preserving the ultrametric inequality is interesting, there is no known practical utility for such a feature. Johnson (1967) also noted that SLINK clustering is invariant to monotonic transformations of the similarity / dissimilarity matrix. Clustering methods are also susceptible to ten types of distortion when representing the structure in a similarity / dissimilarity matrix. Four of these types of distortion will not occur if the cluster analysis method is invariant to monotonic transformation. Invariance under monotonic transformations will make the method most appropriate for use with ordinal data. A visual examination of the dendrogram may not suggest the number of appropriate clusters present in the data.

16.4.2   COMPLETE LINKAGE METHOD (CLINK)

This method is the logical opposite of the SLINK method in that the distance between groups is now defined as the distance between their most remote pair of individuals. This method can be used with both similarity and distance measures. In terms of parameters for recurrent formula, CLINK specifies:

$$A = B = 0.50 \, ; \, C = 0 \, ; \, D = 0.50$$

This method has a tendency to find relatively compact and hyperspherical clusters composed of highly similar cases. A visual examination of the dendrogram gives a clear sense of clusters in the data and this method can be used to indicate the number of appropriate clusters present in the data. The comparison of this cluster membership generated by this method to the known structure may not show high concordance. The CLINK also has the property of being invariant under monotonic transformations of the similarity/dissimilarity measures. In addition, the CLINK does not have the tendency to chain, as in the case of the SLINK method.

The major criticism of CLINK is that it is a space-diluting method. The essence of this criticism lies in the fact that since in CLINK an entity cannot join a cluster until it obtains a given similarity level with all members of a cluster, the probability of a cluster obtaining a new member becomes smaller as the size of the cluster increases. In terms of a multidimensional space, the CLINK is used, as the size of particular cluster increases, the effective distance between the cluster and some nonmember also increases. Hence CLINK dilutes the multidimensional space.

16.4.3   *AVERAGE LINKAGE BETWEEN MERGED GROUPS (ALINKB)*

ALINKB has been proposed as a compromise between the chaining tendency of SLINK and space-diluting tendency of CLINK. Thus Sokal and Sneath called CLINKB as 'unweighted pair-group mean average' (UPGMA) method or simply arithmetic average clustering. To reduce the confusion over cluster analysis terminology, call it as group average method. The distance between the two groups is defined as the average (mean) of the distances of all pairwise combinations between the individuals in the two groups. Again, this method can be used with both the measures of similarity and distances provided the concept of an average is accepted. In terms of parameters of recurrent formula, ALINKB specifies:

$$A = \frac{n_i}{\left(n_i + n_j\right)} \; ; B = \frac{n_j}{\left(n_i + n_j\right)} \; ; C = 0 \; ; D = 0$$

where n_i is defined as the number of entities in cluster i.

Thus, here a cluster is defined as a group of entities in which each member has a greater mean similarity with all members of the same cluster than it does with all members of any other cluster.

16.4.4   *AVERAGE LINKAGE WITHIN THE NEW GROUP (ALINKW)*

In this method, the distance between two groups is defined as the average (mean) of the distances of all pair wise combinations of all individuals in the two groups. If there are n_i entities in the cluster i and if there are n_j entities in cluster j, then there will be $((n_i + n_j)(n_i + n_j - 1))/2$ distinct pair-wise combinations that may be formed from the entities in the merger of i and j. As in the case of ALINKB, this method does not depend on extreme values for defining clusters and accordingly it is not possible to make any statement about the minimum or maximum similarity within a cluser. However, in practice, this method frequently gives results that are little different from those obtained with the complete linkage method. The procedure can be used with both the similarity and distance measures provided the concept of an average measure is accepted.

16.4.5 CENTROID METHOD (CENTROID)

In this method, the groups are depicted to lie in an Euclidean space and are fused according to the distance between their centroids, the groups with the minimum distance are being fused first. Then the groups are replaced on formation by the co-ordinates of their centroids. This procedure is continued till all items form a single group. In terms of parameters for recurrent formula, CENTROID specifies:

$$A = \frac{n_i}{\left(n_i + n_j\right)} ; B = \frac{n_j}{\left(n_i + n_j\right)} ; C = -AB ; D = 0$$

While intuitively appealing, the centroid clustering method is not used much in practice, partly owing to its tendency to produce trees with reversals. Reversals occur when the values at which clusters merge do not increase from one clustering step to the next, but decreases instead. Thus, the tree can collapse onto itself and become difficult for interpretation. The method can be used with both similarity and distance measures.

16.4.6 MEDIAN METHOD (MEDIAN)

This method is a variant of the CENTROID method. There is a disadvantage with the centroid method. If the sizes of two groups to be fused are very different, the centroid of new group will be very close to that of the larger group and may remain within that group. Thus the characteristic properties of the smaller group are then virtually lost. The strategy can be made independent of group sizes by taking the median of the two groups. Thus in the median method, the centroids are weighted equally regardless of how many entities are in the respective clusters. The use of the median can be suggested because when i and j are merged, the distance of a third cluster lies along the median of the triangle that joins the three (i, j, and h)

(A line called the median can be drawn from each vertex of a triangle to the midpoint of the opposite side. The medians intersect in a point.) Here, the distance between clusters is based on the distance between cluster medians. The parameters of the recurrent formula for this method are given by:

$$A = B = 0.50 \; ; \; C = -0.25 \; ; \; D = 0$$

Although, the method could be made suitable for both similarity and distance measures, suggest that it should be regarded as incompatible for measures such as correlation coefficient since interpretation in a geometrical sense is no longer possible.

16.4.7 THE WARD METHOD (WARD)

This method proposes that the loss of information which results from the grouping of individuals into clusters can be measured by the total sum of squares of the deviations of every point from the mean of the cluster to which it belongs. At each step in the analysis, union of every possible pairs of clusters is considered and the two clusters whose fusion results in the minimum increase in the error sum of squares are combined. Thus this method is designed to generate clusters in such a way that the variance within a cluster is minimum. Hence, it is also called as 'Minimum Variance Method'. The parameter values for this method are:

$$A = \frac{(n_i + n_k)}{(n_i + n_j + n_k)} \; ; \; B = \frac{(n_j + n_k)}{(n_i + n_j + n_k)} \frac{n_j}{(n_i + n_j)} \; ; \; C = \frac{-n_k}{(n_i + n_j + n_k)} \; ; \; D = 0$$

The WARD method attracted attention because it is a method which optimizes an objective statistic. Anderberg noted that by minimizing the error sum of squares, the minimum variance method is minimizing tr(W), where W is the pooled within-cluster sum of squares and cross-products matrix. Cormack criticized the WARD method as being biased toward choosing spherical clusters.

This method tends to find clusters of relatively equal size and shape as hyper spheres. A visual examination of the dendrogram from this method gives a sense of clusters. A common problem associated with the use of this method is that the clusters found by this method can be ordered in terms of their overall elevation. It generates solutions that are heavily influenced by profile elevation. Ward method is widely used in many of the social sciences. The ward method is defined only for distance measures.

16.4.8   GENERAL PROBLEMS OF HIERARCHICAL AGGLOMERATIVE METHODS

Hierarchical techniques have no provision for reallocation of entities who may have been poorly classified at an early stage in the analysis. Gowr (1967) gives examples of the way in which early partitions may cut through natural groups.

The fact that there are several hierarchical techniques available, and that several of them can be used with many different similarity coefficients and distance measures, means that there are a number of options open to the investigators.

Hierarchical techniques are probably best suited to biological data for which hierarchical structure can safely be assumed to exist. The hierarchical methods require the calculation and storage of a potentially large similarity matrix. They can generate different solutions simply by reordering the data in the similarity matrix and also they are not stable when cases are dropped out of the analysis.

16.5   NUMBER OF CLUSTERS IN HIERARCHIAL METHODS

The problem of determining the number of clusters present in a data set is common for all methods of cluster analysis.

Determining the number of clusters is an important but often overlooked component in cluster analysis methods. The researcher wants to cluster only when clusters exist and hence it would be useful to have some test for this before going through the whole process of finding best clusters. The ability of certain techniques producing cluster even when clusters were non-existent was well documented. The complexity of the problem of determining the number of clusters, empirical investigations, factors affecting the number of clusters and various analytical procedures to achieve the purpose are briefly narrated in this subsection.

16.5.1   COMPLEXITY OF THE PROBLEM

Among other things, lack of suitable null hypothesis and the complex mature of multivariate sampling distributions are most important reasons causing little progress toward development of methods of determining the number of clusters.

Hierarchical clustering methods give a configuration for every number of clusters from one up to the number of entities in the data. On the other hand, other algorithms find a best fitting structure for a given number of clusters. Some algorithms begin with a chosen number of clusters and then alter this number as indicated by certain criteria with the objective of simultaneously determining both the number of clusters and their configuration.

16.5.2   EMPIRICAL INVESTIGATIONS

The problem of determining the number of clusters is analogous to the problem of determining the number of components in factor analysis. While several empirical investigations have aimed at the evaluation and comparison of the available decision rules on the number of components in factor analysis, such attempts are uncommon in the determination of number of clusters in cluster analysis.

16.5.3   FACTORS AFFECTING NUMBER OF CLUSTERS

The decision on the number of clusters depends on combination of many factors. We usually want the clusters to be a few in number and to be well defined. However, the purpose of the classification is also a factor. Ultimately we want the classification to be judged valid. Thus, we want to get "number of clusters" that make it to work the best. Often it is expected that few well-defined clusters would work better than many clusters.

16.5.4   CLUSTER STRUCTURE AND NUMBER OF CLUSTERS

The first task in cluster analysis is to establish the existence of cluster structure. If this is not established then there is no need to have any cluster analysis for the given data. In the literature, few methods are available to test for the existence of cluster structure.

While few but well-defined methods are available in determining the number of components in factor analysis (Eigen values of one and above,

Scree test, and Horn's Parallel Analysis), numerous methods of determining the number of clusters in the data have been proposed. Some of these methods are based on intuition like examining the dendrogram for large number of clusters between fusions, and the minimum number of entities in the ultimate clusters.

16.5.5 *ANALYTICAL METHODS*

The literature contains a lot of instances for analytical methods of determining the number of clusters. Some authors have been preoccupied with an idea of optimum number of clusters rather than considering the possibility of several alternative classifications, each reflecting a different aspect of the data.

Most of the stopping rules are based on the within and between-cluster distances. Maximizing the between-clusters sum of squares is essentially the criterion used in Linear Discriminant Analysis. The cluster analysts use this criterion to determine the natural clustering of objects from original observations, while the Discriminant Analysis employs it to determine what liner combination of observations best discriminate two a priori clusters.

A simulation study identified six best methods or stopping rules: C-index by Hubert and Levin (1976), Gamm a Index by Baker and Hubert (1975), the point-Biserial index by Mailigan (1980a), G(+) coefficient due to Milligan (1981a), Kendall's Tau by Rohalf (1974) and a Criterion due to McClain and Rao (1975). However, caution is necessary to accept this finding since Milligan and Cooper (1985) stated that the results were based on the type of data, since for a different data, the result would have been different.

16.6 PARTITIONING METHODS

16.6.1 *NATURE OF PARTITIONING METHODS*

The partitioning methods are designed to cluster data units into a single classification of k clusters, where k is either specified a priori or is determined as part of the clustering method. The central idea in most of these methods is to choose some initial partition of the data units and then alter cluster memberships so as to obtain a better partition. Various algorithms which have been proposed differ as to what constitutes a 'better partition' and what methods may be used for achieving improvements. The broad concept for these methods is very similar to that of underlying the steepest descent algorithms used for unconstrained optimization in non linear programming. Such algorithms begin with an initial point and then generate a sequence of moves from one point to another, each giving an improved value of the objective function, until a local optimum is found. Apparently this analogy has not been explored to any great extent; it would be substantial contribution if the accumulated research and experience on unconstrained optimization problem could be used to devise more effective clustering methods. Thus the partioning methods differ from the hierarchical methods. The partitioning methods admit reallocation of the entities and thus allowing poor initial partitions to be corrected at a later stage. Most of these methods employ distinct procedures with respect to the method of initiating clusters and a method of reallocating some or all of the entities to other clusters once an initial classificatory process has been completed.

16.6.2 METHODS OF INITIATING CLUSTERS

There are two basic ways for initiating clusters; one is by using seed points and other by selecting an appropriate starting partition.

a) Starting with Seed Points
The majority of clustering methods begin by finding k-points (seed points) in the p-dimensional space, which acts as initial estimates of the cluster center.

The following methods are representative examples of how such seed points can be generated.

1) Choose the first k data units in the data set. If the initial configuration does not influence the ultimate outcome in any important way, then this method is the simplest.

2) Label the data units from 1 to n and choose those labeled n/k, 2n/k, ..., (k-1)n/k, and n. This method is almost as simple as method 1 but tries to compensate for a natural tendency to arrange the data units in the order of collection or some other nonrandom sequence.

3) Subjectively choose any k data units from the data set.

4) Select k data units at random by using random number tables.

5) Use the centroids of the k clusters of classification obtained by hierarchical methods. Hierarchical grouping ordinarily will make group centroids relatively distinct from each other.

When seed points are used, entities are allocated to the seed points to whose center they are nearest. The seed points remain stationary throughout the assignment of the full data set; consequently, the resulting set of clusters is independent of the sequence in which data units are assigned. The estimates of the cluster center may be updated after the addition of each entity to the cluster. This method bears a strong resemblance to the hierarchical centroid methods described. As in this centroid methods, the clusters centroids migrate so the distance between a given data unit and the centroid of a particular cluster may vary widely during the assignment process; accordingly resulting set of initial clusters is dependent on the order in which data units are assigned.

b) Selecting n Appropriate Starting Partitions

This involves the specification of the first cluster assignment. Various random allocations schemes could be used. For example, divide the sequence of n units into k mutually exclusive clusters at equal or unequal intervals. The difficulty with all such random schemes is that the resulting groups are spread more or less uniformly over the entire data set and their centroids are k different estimates of the data set mean vector. Such groups have no properties of internal homogeneity and are not clusters at all. When starting partitions are used, the centroid of each cluster is defined as the multivariate mean of the cases within the cluster.

16.6.3 METHODS FOR REALLOCATING ENTITIES

The methods for reallocating entities deal with the ways in which cases (entities) are reassigned to clusters. Again, there are two basic types—the k-means passes and the hill climbing passes.

The k-means passes are the commonly partitioning method in many of the social sciences. The k-means passes are also referred to as the 'nearest centroid sorting pass' and the 'reassignment pass'. This simply involves the reassignment of cases to the cluster with the nearest centroid. The k-means passes can be either combinatorial or noncombinatorial. The former method requires the recalculation of the centroid of a cluster after each change in its membership while the later recalculates the cluster centroid only after an entire pass through the data has been completed. One other important distinction is that k-means passes can also be either exclusive or inclusive. Exclusive methods remove the case under consideration from the parent cluster when a centroid is computed, whereas inclusive methods include them. Among the several methods suited to the basic problem of sorting the data units into a fixed number of clusters, the methods suggested by Forgy and a variant proposed by Jancey are very commonly used.

16.6.4   FORGY'S METHOD

Forgy suggested a very simple algorithm consisting of the following sequence of steps:

1) Begin with any desired initial configuration. Go to step 2 if beginning with a set of seed points; go to step 3 if beginning with a partition of the data units.

2) Allocate each data unit to the cluster with the nearest seed point. The seed points remain fixed for a few cycle through the entire data set.

3) Compute new seed points as the centroids of the clusters of data units.

4) Alternate step 2 and 3 until the process converges: that is, continue until no data units change their cluster membership at step 2.

It is not possible to say how many repetitions of steps 2 and 3 will be required to achieve convergence in any particular problem; however, empirical evidence indicates that ordinarily five or less number of repetitions will be sufficient; only infrequently more than 10 repetitions will be needed.

At each repetition, the assignment of a n data units to k clusters require nk distance computations and n(k-1) comparisons of distances. Since k is ordinarily much smaller than n and the number of repetitions to convergence is small, the analyst can often examine sets of clusters associated with several different values of k at less cost than for a full hierarchical analysis.

16.6.5 JANCEY'S VARIANT METHOD

Jancey (1966) independently suggested a method which is similar to Forgy's method but different only at step 3. The first set of cluster seed points is either given or computed as the centroids of cluster in the initial partition; at all succeeding stages each new seed point is found by reflecting the old seed point through the new centroid for the clusters. Since data units were assigned to the cluster on the basis of their proximity to old seed point rather than new seed point, it therefore the new seed points should overshoot the computed centroid. Jancey suggests that this technique will accelerate convergence and possible lead to a better overall solution through bypassing inferior local minima.

Both Jancey's and Forgy's methods implicitly minimize within cluster error function. The cluster boundaries are piecewise linear for both these methods because the boundaries are equidistant from the nearest centroids. Since the seed points are recomputed only after the full data set has been reallocated, the results of these two methods are not affected by the sequence of the data units within the data set.

16.6.6 GENERAL PROBLEMS IN PARTITIONING METHODS

The major problem of partitioning techniques is that sub-optimal solutions are frequently found. This is the problem of local optima. The obvious way to optimize a criterion is to consider every possible partition of the data and choose that one with the optimal value. But in general the number of partitions is very vast, and it is quite impossible to consider every one even by using the most powerful computers available. This has led the development of algorithms designed to search for a local optimum of the criterion.

The partitioning techniques consume large amount of computer time, and are probably not suitable for use with very large data sets, unless a powerful computer is available.

16.6.7 DETERMINATION OF NUMBER OF CLUSTERS

It is generally suggested that a plot (graph) of the criterion value like the point Biserial Correletion against the number of groups will indicate the correct number to consider by showing a sharp increase of the clustering criterion at the correct number of groups.

POINT-BISERIAL CORRELATION

Milligan (1980a) has suggested a simple point-biserial correlation between the raw input dissimilarity matrix and a corresponding matrix consisting of 0 and 1 entries. A value of zero is assigned if two corresponding entries are clustered together by the algorithms and a value of 1 is assigned otherwise. The formula for a point-biserial correlation coefficient reads as,

$$r_{pb} = \frac{(Mp - Mq)\sqrt{pq}}{S.D.}$$

where Mp is the mean dissimilarity of the higher group in the dichotomized, Mq is the mean of the lower group, p is the proportion of cases in the higher group, q is the proportion in the lower group, and SD is the standard deviation of the total samples.

16.6.8   CHOICE OF THE METHOD

It is clear that no one clustering technique can be judged to be the 'best' in all circumstances. In general, several techniques should be used, as this should help to prevent misleading solutions being accepted. This implies the availability of well-documented program listings which can be implemented on the machine available to the user. The method must be compatible with the desired nature of the classification, the variables to be used, and the similarity measure used to estimate the resemblance between cases if one is required.

16.7   VALIDATION TECHNIQUES

Various intuitively reasonable procedures have been suggested for evaluating the stability and the usefulness of the set of clusters obtained. Few methods are reviewed briefly here.

16.7.1 METHODS OF REPLICATION

Since the application of different methods results in different solutions, the same data set may be cluster analyzed by different methods. The solution by the majority of the methods should be similar to say that the data is clearly structured. This involves the estimation of the degree of replicability of a cluster solution by different methods of clustering. If a cluster solution is repeatedly discovered by different procedures on the same set of data, it is possible to conclude that the solution has some generality. A cluster solution that is not stable is unlikely to have general utility. The method of Rand-Index can be used to determine the agreement between solutions by different methods.

RAND INDEX

In an attempt to compare the results of both the hierarchical and the partitioning methods, the problem of assessing the similarity of clusters derived from these methods arose. The rand Index can be used for this purpose. The Rand index is a simple matching coefficient, using 2×2 table, which is a measure of the probability that an entity chosen randomly would have been in the same cluster in two different sets of clusters. In a 2×2 table all possible pair wise combinations, $n(n-1)/2$, are compared with the actual matching from one set of clusters to the other. Thus, the index assumes the form:

$$R = (a+d)/(a+b+c+d)$$

It is noteworthy that both 1-1 and 0-0 matching are incorporated in this index. If all values are 1-1 and 0-0, only cells a and d have any entries, and the index equals 1.0. If all values are in b and c cells the index equals 0.

16.7.2 SIGNIFICANCE TEST ON VARIABLES USED TO CREATE CLUSTERS

A multivariate analysis of variance (MANOVA), multiple ANCOVA, or Chi-Square test on the variables could be used to generate the solution in order to test for the significance of the clusters. The MANOVA, or Ch-

Square techniques can be used on solutions from any clustering technique that create partitions (e.g., Iterative partitioning methods, hierarchical methods etc.). Significance tests for differences among the clusters along these variables should always be positive. Since these tests are positive, regardless of whether clusters exist in the data or not, the performance of these tests is useless at best and misleading at worst.

16.7.3   SIGNIFICANCE TEST ON EXTERNAL VARIABLES

Compare the obtained clusters on other variables of interest which were not included in the original analysis. If difference between clusters persists with respect to these variables, then there is some evidence that a useful solution has been obtained, in the sense that by stating that a particular entity belongs to a particular cluster we convey information on variables not used to produce the clusters. The power of external validity is that it directly tests the generality of a cluster solution against relevant criteria. This approach to validation is not used frequently because of the fact that the methodological design to collect relevant criterion data is usually expensive. A cluster solution that has successfully passed an external validation has much more greater value than a solution that has not.

16.7.4   METHOD OF MARKER SAMPLE

In addition to the methods cited above, one more method for determining the appropriateness of a solution is the method of marker sample. The samples from major clinical diagnostic categories are pooled together as 'marker sample' to aid to validate the cluster solution. A related method in factor analysis is the inclusion of marker variables in the data set to aid in cross-study comparisons. An adequate solution should place dissimilar subjects into different clusters and, additionally, provide some indication of the number of clusters require, to provide reasonable differentiation. For example, it would have been conceptually inconsistent for a given cluster solution to classify numerous psychosomatic and psychotic children together in the same cluster.

16.7.5 ASSOCIATION WITH PRE-PROGRAM ASSESSMENT TEST SCORES

In evaluation of programs, the resultant clusters are evaluated by tests of significance of association with preprogram assessment test scores.

16.8 NUMERICAL DEMONSTRATION

Let us suppose that we have scores of six patients on two diagnostic tests as shown below:

Patient	Test 1	Test 2
A	3	6
B	4	4
C	5	8
D	1	2
E	2	0
F	3	4

The squared Euclidean distance matrix is computed as shown below. For example, the squared Euclidean distance between A and B is given by

$$(3-4)^2 + (6-4)^2 = 1 + 4 = 5.$$

Patient	A	B	C	D	E	F
A	-	5	8	20	37	4
B	-	-	17	13	20	1
C	-	-	-	52	73	20
D	-	-	-	-	5	8
E	-	-	-	-	-	17
F	-	-	-	-	-	-

Let us select complete-linkage method for the purpose of classification by hierarchical method. As a first step in the application, patients B and

F are fused to form a cluster (BF), since the distance within them is the smallest. The distance between the cluster (BF) and the remaining four patients are obtained as shown below.

Distance between (BF) and A = Max (AB, AF)
$$= \text{Max } (5, 4) = 5$$
Distance between (BF) and C = Max (CB, CF)
$$= \text{Max } (17, 20) = 20$$
Distance between (BF) and D = Max (DB, DF)
$$= \text{Max } (13, 8) = 13$$
Distance between (BF) and E = Max (EB, EF)
$$= \text{Max } (20, 17) = 20$$

The distance matrix is given by,

	BF	A	C	D	E
BF	-	5	20	13	20
A	-	-	8	20	37
C	-	-	-	52	73
D	-	-	-	-	5
E	-	-	-	-	-

The smallest entry in the above distance matrix is 5, which is the distance between the patients D and E. Hence, they are fused to form the second cluster (DE). Now the distances are obtained as shown below.

Distance between (DE) and (BF) = Max (D (BF), E (BF))
$$= \text{Max } (13, 20) = 20$$
Distance between (DE) and A = Max (AD, AE)
$$= \text{Max } (20, 37) = 37$$
Distance between (DE) and C = Max (CD, CE)
$$= \text{Max } (52, 73) = 73$$

Now the distance matrix is as given below,

	BF	A	C	(DE)
BF	-	5	20	20
A	-	-	8	37
C	-	-	-	73
DE	-	-	-	-

The smallest entry in the above matrix is 5, which is the distance between (BF) and A. They are fused to form the clusters (BF)A. Now, the distances are obtained as shown below.

Distance between (BE)A and C = Max ((BF)C, AC)
                              = Max (20, 8) = 20
Distance between (BE)A and (DE) = Max ((BF)DE, A(DE))
                                 = Max (20, 37) = 37

Now the distance matrix is as given below,

	(BF)A	C	(DE)
(BF)A	-	20	37
C		-	73
(DE)			-

The smallest entry in the above decision matrix is 20, which is the distance between C and (BF)A. They fused to form the cluster ((BF)A)C. Now the distance is obtained as shown below.

((BFS)C)(DE) = Max ((BFA/C)DE, C(DE)
             = Max (37, 73) = 73

The distance matrix is given below.

	(BFA)C	DE
(BFA)C	-	73
DE	-	-

Finally, the cluster (BFA)C and (DE) are fused at a distance of 73 units to form a single cluster consisting of all the six patients.

The dendrogram is shown in Figure 16.1.

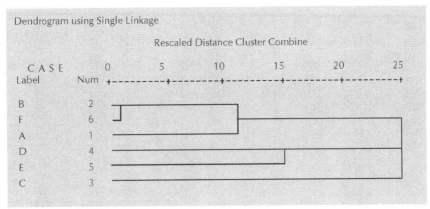

FIGURE 16.1 Dendrogram of six patients based on scores on two diagnostic tests.

16.9 AN EMPIRICAL CLASSIFICATION OF GOVERNMENT MENTAL HOSPITALS IN INDIA

The k-means clustering algorithms are employed in order to cluster 36 government mental hospitals using four key indicators: average bed occupancy, rate of turnover, unit cost (cost per day per patient), and average out-patient attendance. Eight clusters have been obtained. The characteristics and their background information of resulting clusters are as shown in Table 16.1.

TABLE 16.1 Characteristic features of eight clusters of government mental hospitals

Cluster Number	Cluster Name	Number of Hospitals	Average Bed Occupancy	Rate of Turnover	Unit Cost (Rs.)	Average Out-Patient Attendance	Background Information
1	Advanced research institute	1	474	6.7	137	155	Multidisciplinary approach, Research and training, inclusion of community orientation
2	Large size establishments	5	1639	2.3	20	75	High proportion of involuntary patients
3	Out-patient oriented institutes	1	222	5.9	15	320	High follow-up ratio, free voluntary boarding, high unit cost on diet
4	Average hospitals	9	339	4.9	31	93	High proportion of non-functional psychotics and high unit cost on medicine
5	Below average hospitals	8	490	2.7	18	36	High proportion of functional psychotics
6	Hospitals with high turnover rate	7	66	14.4	28	45	Mostly private ownership and voluntary cases

| 7 | Devel-oped hospitals | 4 | 167 | 0.8 | 22 | 10 | Only service facilities for in-voluntary cases. High death rate. High proportion of non-improved cases |
| 8 | Smaller establish-ment | 1 | 16 | 0.6 | 2 | - | Psychiatric wards attached to central jail |

16.10   PATTERN OF DISTRIBUTION OF MENTAL HEALTH MANPOWER: AN INTERNATIONAL SCENE

The data regarding the number of psychiatrists, number of psychologists, number of social workers, and number of psychiatric nurses in 61 nations during 1993 formed the major source of the data for the application of cluster analysis to study the pattern of distribution of mental health man-power. The clusters produced by the Forgy method using the centroids of six clusters of the Ward method are found to be the best classification of the data that was analyzed.

TABLE 16.2   Mental health manpower and background information of six clusters of manpower

Variables	Clusters					
	I	II	III	IV	V	VI
Per one lakh population						
Psychiatrists (1.8)	11.5	6.9	2.2	1.2	0.4	0.1
Psychologists (1.0)	8.6	1.4	0.4	5.5	0.1	0.05
Social workers (2.2)	24.0	1.1	1.3	1.1	0.5	0.03
Psychiatric nurses (9.9)	89.8	31.5	6.1	0.6	0.4	1.5
Background Information						
Per capita income (US$ 4877)	15,513	8,701	6,443	998	910	458
Literacy rate (80%)	99	96	90	85	74	51
Annual growth rate of population (1.9%)	0.6	1.0	1.3	2.4	2.6	3.0

From Table 16.2, it can be noted that the number of psychiatrists per one lakh population was maximum (11.5) in Cluster I, with decrease in subsequent clusters, and it was minimum (0.1) in the last cluster. Thus, these cluster numbers can be considered as grades of these clusters. Except in Cluster IV, the decreasing trend can be seen with respect to the number of psychologists that ranged from 8.6 to 0.05 per one lakh population. There were 5.5 psychologists per one lakh population in Cluster IV. Except in Cluster III, the decreasing trend can be seen with respect to social workers, which ranged from 24.0 to 0.03 per lakh population. There were only 1.1 social workers per one lakh population in Cluster II. Except in Cluster VI, the decreasing trend can be seen with respect to psychiatric nurses, which ranged from 89.8 to 0.4. There were 1.5 psychiatric nurses per thousand populations in Cluster VI.

TABLE 16.3   List of nations in the six clusters

Grade I Cluster	Switzerland	Grade IV Cluster	Dominican Republic
	France		Peru
	Iceland		Colombia
	Norway		Suriname
	Finland		Paraguay
	Denmark		Ecuador
	Israel		Bolivia
	England		
Grade II Cluster	Estonia	Grade V Cluster	United Arab Emirates
	Azerbaijan		Jordan
	Japan		Saudi Arabia
	Turkmenistan		Egypt
	Bulgaria		Lebanon
	Tajikistan		Philippines
	Georgia		St. Lucia
	Hungary		Jamaica
	Ireland		Morocco
	Portugal		Vietnam
	Poland		Thailand

			India
			Haiti
			Kenya
			Ivory Coast
Grade III Cluster	Bahrain	Grade VI Cluster	Indonesia
	Taiwan		Senegal
	Singapore		Nigeria
	South Korea		Zimbabwe
	Malta		Ghana
	Barbados		Zambia
	Trinidad		Namibia
	Mauritius		Nepal
			Tanzania
			Rwanda
			Togo
			Sierre Leone

As shown in Table 16.3, Grade I Cluster consists of 8 nations (Europe: 7, Middle-East: 1), Grade II Cluster consisted of 11 countries (Europe: 6, Asia: 5). Grade III Cluster consisted of 8 countries (Islands: 7, Peninsula: 1), Grade IV Cluster consisted 7 countries (South America-6, an island in the Caribbean Sea: 1), Grade V Cluster consisted of 15 countries (Asia: 8, Africa: 4, islands in the Caribbean Sea: 3), and Grade VI Cluster consisted of 12 countries (Africa: 10, Asia: 2).

16.11   AN EMPIRICAL CLASSIFICATION OF CHILD PSYCHIATRY DISORDERS

16.11.1   FORMAL INTRODUCTION

The urge and ability to classify objects or traits are unique aspects of human experience. It provides us with the capacity to observe, order our observations, and abstract general principles and hypotheses from our experience. Classification enables us to make use of information for purpose of communication, prediction and explanation. At the present time in child and adolescent psychiatry, classification systems have their greatest role in facilitating communication for both clinical and research purposes. The process of assigning a label may itself be associated with some sense of relief on the part of the patient, or the patient's parents. Sometimes this reflects the mistaken notion that having a label implies having an explanation. Although the classification of psychiatric disorders may be considered as 'labeling people' and lead to harmful effects, the classification should not be avoided, but it should be strengthened.

16.11.2 CRITERIA FOR CLASSIFICATION

The classification systems vary depending upon the purpose of classification and what is being classified. The classification scheme must easily and reliably be used by clinicians and researchers, and hence the need for phenomenological based, readily compressible diagnostic terms and descriptions. The childhood mental disorder could be described so that they can be differentiated from one another. They should differ in important respects, such as associated features and courses. The classification system must be applicable over the range of development and must be comprehensive and logically consistent. General classification system must cover the entire range of disorders in a logically consistent fashion; classification systems developed for a high specific purpose do not share this concern. The need for reasonable parsimony must be balanced with the need for adequate coverage.

16.11.3 CLASSIFICATION OF DISORDERS AND CLASSIFICATION OF CHILDREN

Opinion differs among the scientists and it is often debated whether to classify disorders or children. It is important to classify disorders, rather than children. Concern has been raised about the possible effects of labeling children, and to some extent this concern is valid. It is, of course, also the case that having an adequate label for a child's disorder may be helpful in securing him or her needed services. In this regard it is always important to refer to the child's disorder, not to the child as the disorder. The term diagnosis refers both to the notion of assigning a label to a given problem and to the act of evaluation. In important respects it is the diagnostic process that is most important of the two.

Although diagnostic label have considerable value, they do not provide information about the individual, which is unique and uniquely related to intervention. Diagnostic categories will and should, change, and children may exhibit a disorder for variable periods of time. The need of individuals will vary depending on the individual and not simply as a function of whatever disorders he or she has.

16.11.4   ISSUES IN CLASSIFICATION

There are several issues involved in the classification of child and adolescent psychiatry. Developmental considerations assume major importance in the provision of a psychiatric scheme for children, and indeed, for adults as well. Theoretical models of psychological disturbances have been developed from rather diverse historical traditions; they have considerable value for the individual clinician in understanding and treating children with emotional and behavioral problems. Theory-driven classification schemes are, by their nature, based on a set of assumptions and hypotheses that are not generally shared. Classification systems may be developed to approximate some ideal diagnostic system in which etiology could be directly related to clinical condition. But different etiological factors may result in rather similar conditions, and the same etiological factors may be associated with a range of clinical conditions. Aspect of intervention may be more directly related to the clinical condition than to etiology. In certain situations and populations, contextual variables such as family, school, or cultural settings pose major complications for application of diagnostic systems.

Issues of reliability and validity remain to be addressed for many categories and classification systems. The attempt to address these issues through empirical data rather than theorizing is, perhaps, the greatest accomplishment of the past 15 years.

16.11.5   CLASSIFICATION MODELS

Three different approaches to classification of disorders have been identified; the ideographic, the categorical, and the dimensional.

16.11.5.1   IDEOGRAPHIC APPROACH

This approach focuses on the total context of the child's life in formulating a case; these classifications may be theory driven (eg. by psychoanalytic or behavioral theories) or may be used eclectically. Ideographic approaches are commonly used in clinical work. In this approach, the child or adolescent is viewed in the totality of his or her life circumstances, various disorders, and problems. Psychosocial situations may be viewed as worthy of notation and treatment. These approaches make it very difficult to communicate information for clinical and research purposes in a concise and readily understandable fashion.

16.11.5.2 CATEGORICAL APPROACH

The categorical approach is referred to as the medical model of classification which views disorders as either present or absent. This system generally begins when individuals set up clinical criteria for a particular diagnosis. Based on the diagnostic evaluation, it is determined whether or not the patient's disorder meets the criteria for one or more of these diagnostic categories.

The categorical approach owes much to the work of clinical investigators at Washington University in St. Louis. They proposed a five-stage model for adult psychiatric classification. Cantwell modified this five-stage model into a six-stage model for valid psychiatric disorders of childhood. The first five stages are the clinical description, physical and neurological factors, laboratory studies, the family studies, and natural history studies. The sixth phase of his model is treatment studies, based on the hypotheses that marked differences in respond to treatment, such as complete versus marked deterioration, should be considered evidence that the original group of children did not form a homogeneous group. Differential response could be used to subdivide the original index population of patients. An example of this might be children with attention deficit disorder who do and do not have a dramatic response to stimulant medication. It is appeared that these six stages interact with one another and that new findings in one stage may lead to changes in one or more of the other stages.

The 'official' psychiatric classification systems, viz: the Diagnostic and Statistical Manuals of Mental Disorders (DSM) of the American Psychiatric Association (APA) and the International Classification of Diseases (ICD) of the World Health Organization (WHO), have generally been categorical. All the evidences of DSM reveal the increasing professional attention paid to child and adolescent psychopathology. These manuals, in part, include the changing conceptualizations of maladaptive behavior for these age groups. This traditional diagnostic system has been challenged on the basis of reliability and validity of the clinical syndromes offered. Although the third edition of DSM requires that specific criteria be present before a diagnosis can be made, field studies demonstrated reasonable inter-rater reliability for only broad diagnostic grouping. The more specific diagnostic categories that are used in clinical practice obtained limited evidence of such reliability. In addition, questions have been raised as to whether sufficient evidence exists to justify the inclusion of certain diagnoses. Garmezy concluded that few diagnostic categories had been adequately validated.

16.11.5.3   DIMENSIONAL APPROACHES

Unlike the categorical approach which views disorders as dichotomous, the dimensional approach to classification relies on assessment of dimensions of function or dysfunction by reducing phenomena to various dimensions along which a child can be placed. Various sources of data can be used for this approach, such as behavioral ratings, parental reports, yes/no criteria, developmentally based test scores, and the like. Although the dimensional approach is more commonly used in nonmedical settings, many medical phenomena also exhibit continuous (i.e., dimensional) characteristics. For some purposes categorical diagnoses (e.g., level of mental retardation) are derived from what is essentially a continuous variable, while some dimensional assessment instruments can similarly be used to generate categorical diagnoses. Various statistical techniques such as factor and cluster analysis may be used to derive relevant clinical patterns or profiles. These patterns may, in turn, be used to derive categorical diagnoses.

Consistent findings in empirical studies have provided the impetus for this approach. Two comprehensive reviews indicated relatively consistent factor, analytically derived dimensions of psychopathology. Variation among studies in terms of number and content of the factors relates to behaviors assessed, population studies, factor algorithms, and the investigator's conceptual framework. Overall, similar factor dimensions are extracted across populations, methods of data collection, and source of this data (e.g., parent, clinician, and teacher). The relative invariance of the dimensions has led to investigation in which they are used in the development of empirically desired classification schemes.

16.11.6   APPLICATIONS ON REAL DATA

Six clusters of patients obtained by employing k-means algorithms of the Forgy method using the centroids generated by complete-linkage and

Euclidean distances, clustering technique on 435 child guidance clinic cases were obtained. They were named as childhood psychosis, hysterical syndrome, anxiety disorders, conduct disorders, hyperactivity syndrome and scholastic backwardness. These clusters were characterized by the frequency of occurrence of all the variables included in the study as shown in Appendix-IV. The clusters have good communicative values for both research and clinical work, reflect actual clinical experience and further may suggest an ideal diagnostic classification system.

16.12   CLUSTER ANALYSIS IN EVALUATION OF PROGRAMS

16.12.1   A FORMAL INTRODUCTION

Evaluation study is made for assessing the effectiveness of program implemented or for assessing the impact of developmental projects on the development of project area. Evaluation is determination of the results attained by some activity (whether a program, a drop or a therapy or an approach) designed to accomplish some valued goal or objective. Evaluation research is thus, directed to assess or appraise the quality and quantity of an activities and its performance, and to specify its attributes and conditions required for its success. It is also concerned with change over time. As evaluation research asks about the kind of change the program views as desirable, the means by which the change is to be brought about, and the signs according to which such change can be recognized.

16.12.2   EVALUATION APPROACHES

These are different kind of approaches to evaluation, based on when it occurs (process vs. outcome), the intensions of the evaluator (formative vs. summative), and who does the evaluation (internal vs. external).

16.12.2.1   PROCESS AND OUTCOME EVALUATION

Evaluations may be conducted at any of the several phases of program development and implementation. Process and outcome evaluations relate to the phase of the program studied. Process evaluations examine what goes on inside the program while it is in progress, focusing on such things as activities offered, staff practices, and client actions. Evaluating program processes may be very helpful in understanding why program objectives are or are not met. Outcome evaluations put the emphasis on the program objectives or outcomes: What happens to clients as a result of their participation in the program. Both process and outcome evaluations may be either formative or summative, depending upon the intentions of the evaluator (or those who hire the evaluator).

16.12.2.2   FORMATIVE AND SUMMATIVE EVALUATION

Program sponsors and tax payers typically are interested in program accountability, which calls for summative evaluation. Summative evaluation helps to answer the question, 'Is the program earning its keep?' and provides information which can be used to help program sponsors make informed decisions about program funding. In comparison, people who implement programs are often more interested in learning how to improve their programs. Program improvement is the purpose of formative evaluation.

16.12.2.3   INTERNAL AND EXTERNAL EVALUATION

When an organization has an evaluator or evaluation team on staff to evaluate their own programs, this is considered internal evaluation. External evaluations are conducted by outside evaluation organizations or independent evaluators. Some factors which should be considered in deciding whether to do an internal or external evaluation are public confidence in the results, the likelihood that the evaluation will be objective, the evaluator's depth of understanding of the program, and the potential for utilization of the results.

16.12.3   TYPES OF EVALUATION

Evaluations are of three types.

16.12.3.1 CONCURRENT EVALUATION

Concurrent evaluation is a continuous process and partakes the nature of an inspection or social audit of an on-going program. It aims at the evaluation of the quality implementation and services as a feed back for improving the performance.

16.12.3.2 PERIODIC EVALUATION

Periodic evaluation is made after each distinct phase or state of a project has been completed. In case of a medium period time-bound program like 5 years plan, this evaluation may be done in the middle of the period and it may be called mid-term or interim evaluation.

16.12.3.3 TERMINAL EVALUATION

Terminal evaluation is done after the completion of a program or project. This is designed to assess the extent of the achievement of its goals or objectives. It may also involve a benefit-cost analysis. In case of a project with long gestation period, the appropriate methodology for terminal evaluation will consists of survey-cum-experimental design.

16.12.4 EVALUATION RESEARCH AND OTHER TYPES OF RESEARCH

16.12.4.1 Distinction Between Evaluation Research and Other Types of Research

There are a number of specific criteria which distinguish evaluation research from other types of research. Evaluation research is wholly conducted for a client who intends to use the finding as a basis of decision making. This is quite different from basic research, which aims at knowledge for its own sake. The evaluation researcher deals with his clients questions relating to the latter's program, while the basic researcher formulates its own research questions. The evaluation researcher measures whether the program goals are being reached. Unlike basic researcher who normally has control over research work, the evaluation researcher works in a setting where priority goes to the program as opposed to the evaluation. This means that the evaluation researcher must fit time schedule to the program built-in time schedule. The program staff tends to see data-collection work as hindrance to their work. Researcher-program personnel conflicts are inherent in evaluation study. While researcher is interested in objective evaluation and public dissemination of results, the project personnel expect that the evaluation results should be meant for in-house use only.

16.12.4.2 Similarity between Evaluation Research and Other Types of Research

Evaluation research does not differ from other types of research in methodology. The problems of reliability, validity and operationalization and research method, techniques and principles are common to evaluation and other types of research. An evaluation researcher must plan his design as to probe deep into questions such as.

1. What is the nature of content of the objectives of program to be evaluated
2. Are the objectives unitary or multiple
3. How are the objectives to be achieved
4. Who is the target of the program
5. When is the desired change to take place
6. What is the desired magnitude of effect
7. Do the benefits really reach the target group
8. What are the unintentional effects or side effects of the program

The evaluation researcher's operationalization is concerned with indicators of input activity variables, program output in terms of the policy objectives, adequacy of performance (one common index of adequacy is the rate of effectiveness divided by the number of persons exposed to the program), efficiency to measure of benefit-cost, and process to measure why the program is successful or unsuccessful.

16.12.5 EVALUATION AND RESEARCH

Evaluation is systematic inquiry designed to provide information to decision makers and other parties interested in a particular program, policy, or intervention. In defining evaluation, it may be helpful to compare it to research. Both research and evaluation are based on systematic inquiry. But while evaluation makes use of research methodologies and designs, evaluation differs from research in a number of ways. Two key distinctions lie in the knowledge base of the researcher or evaluator and his or her purpose in conducting systematic inquiry. One might think of research as aimed at 'knowing how' something works and evaluation as 'knowing how well' something works. To conduct quality research in reading strategies, for example, one must be well grounded in the reading process and be aware of the unanswered questions about how it works. An evaluator can determine the effectiveness of a reading strategy, however, without an in-depth understanding of how it works. Instead, the evaluator wants to know if it works, how well it works, and perhaps, how it was implemented. In other words, evaluation tends to be concerned with quality. The evaluator compares the operation and outcomes of a program to a set of explicit or implicit standards. Decision makers may then use these comparisons to make judgments' about the worth of a program.

16.12.6 *NEEDS ASSESSMENT AND PROGRAM PLANNING*

Most people think of evaluation as something that happens once a program has been in place for a while. But evaluation is an important component in designing and planning programs as well. A critical element of effective program planning is something called 'needs assessment'. In conducting a needs assessment, we identify and measure the level of unmet needs in programs, organizations, or communities. We define needs as the things that people must have in order to be in a satisfactory or describe state. Educational, psychological, and social needs are very complex, and these needs often are not fully understood. So that we may better understand these complex needs, we seek relevant information from a variety of sources. Sources of information may include community or organization surveys, in-depth interviews, expert informants, focus groups, social indicators, and/or large-scale databases. Because many program providers and planners are not data-oriented, comfortable in constructing surveys, or able to analyze the information obtained in such surveys and other data sources, evaluators need to provide these valuable services to program planners and providers. Evaluators can also raise questions about assumptions implicit in proposed interventions and the steps involved in implementing the program, thus helping to clarify precisely what it is the program hopes to accomplish and the activities that will most likely be effective in achieving those goals.

16.12.7 POTENTIAL OF EVALUATION

For many individuals and organizations, the term evaluation has negative connotations. Most of us really don't like the idea of being judged or having our shortcomings pointed out by someone else. Almost all of us get nervous at the mere thought of being watched and having our performance assessed for the purpose of evaluation. It gets worse if we know that the evaluation is high-stakes, that is, if something (like a raise in pay or a demotion) is dependent on the outcome of the evaluation process. Some of the formal evaluations of programs that have taken place throughout the history of evaluation have not done anything to allay our fears. The focus of early formal evaluation studies was summative evaluation, with the purpose of helping program sponsors to decide whether or not the program should continue to receive funding. Sometimes these evaluations were technically adequate and fair and sometimes they were not. Today, the focus has shifted to formative evaluations for the purpose of helping to improve programs and components within programs. There is also more recognition of the complexity of program delivery and what constitutes success, so that more sources of information and more points of view are considered in the evaluation process. Evaluators can legitimately emphasize help and assistance to programs, while still contributing to program accountability.

So today, evaluation has tremendous potential for helping us to make positive changes in our own lives and in the lives of our programs and organizations. At PREL, we believe that careful, fair, and diplomatic evaluation studies will provide program planners and providers with information that is critical to success in meeting their short-term objectives and long-term goals. Evaluation is a necessary to the accountability of programs for which the public clamors. It is also necessary to the understanding that must be developed for us to learn why something works, so that we may continue to improve. A better understanding of the importance, roles, strengths, and limitations of evaluation methods on the part of program sponsors, planners, providers, and participants will help evaluators to deliver on the promise of evaluation.

16.12.8 STATISTICAL METHODS IN EVALUATION OF PROGRAMS

The statistical methods in evaluation of programs in pre- and post-program assessment data are as follows:

1. McNemar's test for nominal scale data
2. Sign test and Signed-rank test for ranks scale data (ordinal data) and
3. Paired *t*-test for interval scale data

The statistical methods in evaluation of programs used in repeated measure' data are as follows:

1. Two-way ANOVA based on Ranks
2. Repeated observation ANOVA

The use of analysis of covariance methods may not directly evaluate programs, but can be used in determining whether the mean post-test scores of several variables / groups can be considered as equal.

The multivariate analysis of covariance (MANCOVA) is an extension of ANCOVA. It is simply MANOVA where the artificial dependent variables are initially adjusted for difference in one or more covariates.

16.12.9 CLUSTERING APPROACH IN EVALUATION OF SCHOLASTIC IMPROVEMENT PROGRAM OF CHILDREN

16.12.9.1 NEED FOR CLUSTERING APPROACH

There is a dearth of application of cluster analyzed methods in the field of evaluation research. Those who employ univariate methods were discontent with their final results of determining the pattern of improvements of programs.

16.12.9.2 METHODS

It was proposed to employ multivariate statistical techniques of cluster analysis and standardized difference scores obtained from pre- and post-assessment of study variables to determine the pattern of scholastic improvement in the rural children and validate the results by correlating with demographic and psychosocial variables as well as pre-assessment scores.

16.12.9.3 MATERIALS

The study involved in the administration of educational tests on five dimensions, viz. colored progressive matrices for intelligence, attention, arithmetic, vocabulary, and creativity. The data was collected once before giving the training and once after giving the training program for promotion of these areas. The data was also available on the social and demographic information about the children, and behavioral, emotional and psychological information derived from the Rutter scale. After the pre-assessment of academic tests 24 sessions of interventions have been introduced. The relevant data was collected once before giving the training and once after giving the training program for promotion of these areas. The following interventional program were given to assess the scholastic performance of children and consists of five main objectives, viz. a) Art and craft work, b) Games and play activities, c) Word and vocabulary games, d) Cultural activities, and e) Number games and miscellaneous activities.

Only children from class 3 to class 7 and aged between 7 and 15 years were included for the sake of uniformity for evaluation tests employed on the children to assess the scholastic performance. After fulfilling the above criteria, only 474 children were obtained whose data was available and complete in all relevant aspects.

16.12.9.4 PRELIMINARY ANALYSIS

As a preliminary analysis of the effectiveness of the scholastic improvement program, the pre- and post-assessment mean scores, their differences and effect sizes for all the five dimensions of scholastic performance of 474 school children were determined. As shown in Table 16.1, in the ntelligence dimension, the mean test score was 14.5, which increased to 17.3 in the post-intervention. In attention dimension, the mean pre-test score was 56.9 has increased to 77.6 in the post-intervention. In arithmetic dimension, the mean pre-test score was 17.5, which has increased to 19.4 in the post intervention. In vocabulary dimension, the mean pre-test score was 11.8, which has increased to 17.8 in the post-intervention. In creativity dimension, the mean pre-test score was 15.8, which has increased to 27.3 in the post-intervention.

16.12.9.5   CLUSTERING METHODS

By executing the program SPSS and difference scores of each measured on 22 continuous variables, 21 dendrograms (seve methods of clustering each with three distance measures) were obtained. The K-means with the centroids of CLINK-Manhattan has the highest number of statistical criteria and hence it was selected as the best method.

The best pattern finally selected by the present study consisted of 5 clusters. As shown in Table 16.4, the first cluster consisted of 32 children and they have significant improvement in all the levels of intelligence and hence they named as intelligence improvement. Further, as shown in Table 16.5 and 16.6, they have also characterized by high proportion of females, and low percentage of all types of psychological problems. The second cluster consists of 106 children and they have significant improvement in all levels of attention. Further, they are characterized by significantly high percentage of hyper activity, developmental and neurotic problems. The cluster is named as attention improvement.

The third cluster consists of 186 children and they have significant improvement in all the levels of mathematic and creativity and hence named as creativity and arithmetic improvement. The cluster is characterized by the presence of low percentage of hyper activity and behavioral problems.

The fourth cluster consisted of 130 children and they have significant improvement in all levels of vocabulary. The cluster is named as vocabulary improvement. Further, they are characterized by older children and higher educational standard and from tribal families. They were further characterized by the presence of school-related problems.

The fifth cluster consisted of 20 children and they have significant improvement in all the five dimensions of scholastic improvement program. But, the pre-test assessment scores indicated that they have improved in all the dimensions, and hence they have named as already improved. Further, they have characterized by male, tribal, 6th standard education class and older children in the age group of 12 to 15 years. They were further characterized by the high percentage of writing difficulties.

TABLE 16.4 Pre, Post- and difference scores on five dimensions of scholastic performance in five clusters of children

Dimension	All Children (474)	Cluster 1 (32)	Cluster 2 (106)	Cluster 3 (186)	Cluster 4 (130)	Cluster 5 (20)
Pre-test scores						
1. Intelligence	14.5	14.3	14.5	13.6	14.3	25.3*
2. Attention	56.9	56.8	50.4	55.8	58.8	88.6*
3. Arithmetic	17.5	19.0	17.1	14.6	19.3	32.3*
4. Vocabulary	11.8	13.3	10.9	10.7	10.8	30.0*
5. Creativity	18.8	19.7	16.8	13.9	15.8	21.6*
Post-test scores						
1. Intelligence	17.3	25.8*	16.7	15.0	17.7	26.3*
2. Attention	77.6	81.6	106.2*	60.3	75.9	90.4*
3. Arithmetic	19.4	19.4	22.3	28.4*	21.7	33.5*
4. Vocabulary	17.8	12.0	14.0	13.7	26.2*	31.4*
5. Creativity	27.3	18.8	20.5	36.6*	22.1	23.1
Difference scores						
1. Intelligence	2.8	11.5*	2.2	1.4	3.4	1.0
2. Attention	20.7	24.8	55.8*	4.5	17.1	1.8
3. Arithmetic	7.3	0.4	5.2	13.8*	2.4	1.2
4. Vocabulary	6.1	-1.3	3.1	3.0	15.4*	1.4
5. Creativity	11.5	-0.9	3.7	22.7*	6.3	1.4

TABLE 16.5   Socio-demographic characteristics of five clusters of children studied (figures in %)

Variable	Total (474)	Cluster 1 Intelligence Improvement (32)	Cluster 2 Attention Improve- ment (106)	Cluster 3 Cre- ativity and Arithmetic Improvement (186)	Cluster 4 Vocabulary Improve- ment (130)	Cluster 5 Already Improved (20)
1. Age						
7-9 years	27.2	-	-	-	-	-
10-12 years	63.5	-	-	-	-	-
13-15 years	11.6	-	-	-	20.0	20.0
2. Gender						
Male	46.2	-	-	-	-	65.0
Female	53.8	78.1	-	-	-	-
3. Class						
III Standard	15.4	-	-	-	-	-
IV Standard	24.3	-	-	-	-	-
V Standard	30.4	-	-	-	-	-
VI Standard	22.8	-	-	-	-	40.0
VII Standard	13.6	-	-	-	23.8	-
4. Family Income per month						
Below Rs. 2,500	29.1	-	-	-	-	-
Above Rs. 2,501	70.9	-	-	80.1	-	-
5. Caste						
Tribal	45.6	-	-	-	58.5	75.0
Non-tribal	54.4	-	-	75.8	-	-

Only significantally high percentage figures are given

TABLE 16.6   Behavioral, emotional, social and school related problems of five clusters of children studied (figures in %)

Variables	Total (474)	Cluster 1 Intelligence Improvement (32)	Cluster 2 Attention Improvement (106)	Cluster 3 Creativity and Arithmetic Improvement (186)	Cluster 4 Vocabulary Improvement (130)	Cluster 5 Already Improved (20)
1. Poor performance in the examination	24.6	-	-	-	94.6	-
2. Inconsistence academic performance	43.7	-	-	-	50.8	-
3. Poor attendance in school	2.2	-	-	-	4.6	-
4. Truancy from school	15.9	-	-	-	23.1	-
5. Trend to be absent from school for trivial reason	14.9	-	-	-	-	-
6. Nick name in the school	14.5	-	-	-	-	-
7. Not much liked by other children	5.4	-	-	-	-	-
8. Has tears on arrival of school	4.1	-	-	-	-	-
9. Wet / soiled in the school	3.8	-	-	-	-	-
10. Stammering/stuttering	3.2	-	7.5	-	-	-
11. Other speech difficulty	3.4	-	7.5	-	-	-
12. Specific reading difficulty	10.2	-	17.9	-	-	-

TABLE 16.6 *(Continued)*

13. Specific writing difficulty	16.3	-	25.5	-	-	30.0
14. Often destroys own/ other belongings	4.5	-	-	-	-	-
15. Frequently fights with other children	7.0	-	-	-	-	-
16. Often tells lies	11.8	-	-	-	-	-
17. Stolen things once/ more occasions	5.0	-	-	-	8.5	-
18. Bullying other children	12.7	-	-	-	17.7	-
19. Often dis-obedience	10.2	-	-	-		-
20. Restless/ hardly ever still	35.8	-	53.8	-	44.6	--
21. Poor con-centration/ short atten-tion span	11.6	-	15.3	-	16.9	-
22. Fussy/ over particular child	31.9	-	48.1	-	-	-
23. Squiring fidgety child	14.9	-	22.6	-	20.8	-
24. Do things on his own	11.6	-	17.9	-	16.9	-
25. Often worried / worries	6.6	-	-	-	-	-

TABLE 16.6 *(Continued)*

26. Irritable	22.4	-	-	-	-	-
27. Often appears miserable	14.9	-	-	-	-	-
28. Has twitches/ mannerism	6.5	-	11.3	-	-	-
29. Frequently sucks thumbs / fingers	4.5	-	8.5	-	-	-
30. Often bites nails/ fingers	8.4	-	13.2	-	-	-
31. Tends to be tearful / afraid of new things	18.8	-	-	-	-	-
32. Any physi- cal handi- caps	7.9	-	-	-	-	-
33. Often com- plaints pain or ache	6.8	-	-	-	-	-
34. Extra curricular activities	67.9	-	-	-	-	-
35. Psychologi- cal help	1.1	-	-	-	-	5.0

KEYWORDS

- **Cluster analysis**
- **Dendrogram**
- **Jaccard's coefficient**

CHAPTER 17

DISCRIMINANT ANALYSIS

CONTENTS

17.1 TWO-GROUP DISCRIMINANT ANALYSIS

Here, we are provided with two classes (groups) of persons, and t measurements have been made on each person.

17.1.1 VECTOR OF WEIGHTS

We aim at finding out the weighing vector that when applied to some newly observed and unclassified person will assign him to one or other of the classes with the smallest probability of error. The vector of weights (w) which provides the optimum assignment is given by the following relation:

$$w = V^{-1} d$$

where V is the weighted average of the variance–covariance matrix. It is given by the following relation:

$$V = \frac{W_1 + W_2}{(n_1 + n_2 - 2)}$$

where d is the vector of differences between the t-pairs of means of the two groups.

17.1.2 CUT-OFF SCORE FOR CLASSIFICATION

Generally, for classification purpose, the appropriate cut-off point score is the half-way mark between the means of the discriminant scores of the two groups.

17.1.3 PROBABILITY OF MISCLASSIFICATION

The probability of misclassification is given by the following relation:

$$P(\text{mc}) = 0.5 - P\,(0.5\ D)$$

where D^2 is the generalized distance (Mahalanobis D^2) between two groups. It is the inner product of the vector of difference d and the vector of weights w. That is,

$$D^2 = d'w = d' V^{-1} d$$

It can be shown that the D^2 is the difference between the means of the two groups on discriminant function scores. In addition it can be shown that half of D is a unit normal deviate. Hence, it can be used to determine the probability of misclassification. If we look it up in a table of Z-distribution, we find the probability of an individual who really belongs to one group will be incorrectly assigned to the other group by the discriminant function analysis.

17.2   K-GROUP DISCRIMINANT ANALYSIS

The technique of the linear discriminate analysis developed for the two groups can be extended to the case of more than two groups.

17.2.1   WEIGHT OF UNCLASSIFIED PERSON IN VARIOUS GROUPS

The weight of the unclassified person (undiagnosed patient) in the ith group (diagnosis) is given by the following relation:

$$w_i = p'l_i - \tfrac{1}{2} m_i' l_i$$

where p is the column vector of scores of the unclassified person. The l_i is the vector of weights for the ith group. It is given by

$$l_i = V^{-1} m_i$$

m_i is the vector of means of the ith group
    The procedure is to assign the person to the group for which he obtains the highest score.

17.2.2 NUMERICAL DEMONSTRATION

Let us suppose that a child psychiatrist has to take a decision whether conduct disorder or hyperkinetic diagnosis may be more suitable for a child registered at a child guidance clinic. The information on which the decision is based on is derived from the scores of three genuine conduct disorder children and three genuine hyperkinetic children as the basis for our advice. The children obtained the following scores on two diagnostic tests:

Conduct Disorder Child	Test 1	Test 2
C1	3	6
C2	4	4
C3	5	8
Mean	4	6

Hyperkinetic Child	Test 1	Test 2
C4	1	2
C5	2	0
C6	3	4
Mean	2	2

We calculated the sum of squares and sum of products matrix of each diagnostic groups.

$$W_1 = \begin{bmatrix} 2 & 2 \\ 2 & 8 \end{bmatrix} W_2 = \begin{bmatrix} 2 & 2 \\ 2 & 8 \end{bmatrix} W = \begin{bmatrix} 4 & 4 \\ 4 & 16 \end{bmatrix}$$

The weighted average of the variance–covariance matrix is as given below:

$$V = \begin{bmatrix} 4 & 4 \\ 4 & 16 \end{bmatrix} \Big/ 4 = \begin{bmatrix} 1 & 1 \\ 1 & 4 \end{bmatrix}$$

And $V^{-1} = \begin{bmatrix} \dfrac{4}{3} & \dfrac{-1}{3} \\ \dfrac{-1}{3} & \dfrac{1}{3} \end{bmatrix}$ and $d = \begin{pmatrix} 2 \\ 4 \end{pmatrix}$

Hence, $w = \begin{pmatrix} \dfrac{4}{3} \\ \dfrac{2}{3} \end{pmatrix}$

Now, the discriminate scores of the six children are calculated as shown in the following table. For example, the discriminate scores of the first child are calculated as follows:

$$\left(\frac{4}{3} \times 3 \right) + \left(\frac{2}{3} \times 6 \right) = 8$$

Conduct Disorder Child	Discriminate Score
C1	8.00
C2	8.00
C3	12.00
Mean	9.33

Hyperkinetic Child	Discriminate Score
C4	2.67
C5	2.67
C6	6.67
Mean	4.00

The half-way mark (cut-off score) is given by

$$(9.33 + 4.00)/2 = 6.67$$

Let we support that the scores of an undiagnosed child are given by 2 and 5 on test 1 and 2, respectively. The discriminate scores of this child is calculated as,

$$(2 \times 4/3) + (5 \times 2/3) = (8/3 + 10/3) = 6.00$$

Since, this value is less than the cut-off score of 6.67, the child is most appropriately diagnosed as hyperkinetic child.

The generalized distance (D^2) is calculated as follows:

$$D^2 = (2 \quad 4) \begin{pmatrix} \dfrac{4}{3} \\ \dfrac{2}{3} \end{pmatrix}$$

$$= (8/3 + 8/3) = 16/3 = 5.33$$

D^2 can also be obtained as follows:

D^2 = (mean of the first group) − (mean of the second group)

$= 9.33 - 4.00 = 5.33$

$D = 2.31$

Now,

$P(mc) = 0.50 - P(0.5 \times 2.31) = 0.50 - P(1.155)$

$= 0.05 - 0.37 = 0.13$

Thus, the probability of misclassification is 13 percent that means that we may expect 87 percent of our assignments (diagnosis) to be correct.

17.3   A MODEL FOR DIFFERENTIATING CONDUCT DISORDER AND HYPERACTIVITY IN CHILDREN

A method of linear discriminant analysis of 24 conduct disorder and 16 hyperactivity children, each were measured on 29 symptoms of binary in nature in order to develop a model for differentiating conduct disorder and hyperactivity in children.

TABLE 17.1   Percentage of occurrence of discriminating symptoms in two groups of children studied

Symptoms		Conduct Disorder (24)	Hyperactivity (16)	Significance of Difference P-value
Developmental History				
1	Postnatal problem of child	-	25	0.02
2	Speech / language	13	50	0.02
3	Emotional-social	-	50	0.01
Developmental Problems				
4	Speech / articulation	17	50	0.03
5	Language (expressive)	13	44	0.03
6	Inability to relate to people	13	50	0.01
7	Feeding problems	4	50	0.01
Psychopathology				
8	Poor attention	50	100	0.01
9	Distractible	33	100	0.01
10	Restless	33	100	0.01
11	Impulsive	25	94	0.01
12	Stubborn	83	50	0.03
13	Disobedient	92	50	0.01
14	Quarrelsome	79	31	0.01
15	Aggressive	88	50	0.01
16	Temper tantrum	83	6	0.01
17	Truancy	67	13	0.01
18	Lying / stealing / cheating	63	6	0.01
19	School refusal	71	19	0.01
Psychosocial Stressors				
20	Family history of mental retardation	4	31	0.03
21	Problems with parents	42	13	0.05
22	Marital disharmony	54	13	0.01
23	Change in school/medium subjects	42	13	0.05
Temperamental Dimensions				
24	Dependability	33	75	0.01
25	Sensitivity (self)	75	38	0.02
26	Abnormal activity	63	94	0.03
27	Not social (family)	42	13	0.05
28	Aggressive (verbal)	88	19	0.01
Social Support				
29	Having friends	33	6	0.05

In the first application of the method, 10 symptoms lead to apparently obscured results since their contributions to discrimination has negative sign. The symptoms with the maximum negative contribution were excluded, and the LDFA was carried out on the number of symptoms. The step-wise procedure was repeated until all the resultant symptoms had positive contributions. The vector of weights and the vector of contribution in the final application with 12 symptoms are presented in Table 17.2

TABLE 17.2   Vector of weights and vector of percentage contribution of 12 symptoms

	Symptoms	Weights	Percentage Contribution Added
1	Postnatal problems of child	-5.60	4.8
2	Family history of mental retardation	-7.81	2.8
3	Poor attention	-7.50	12.8
4	Problems in language expression	-1.72	1.8
5	Impulsive	-5.43	12.7
6	Restless	-0.25	0.3
7	Temper tantrum	11.44	30.1
8	Lying / cheating / stealing	4.72	9.1
9	School refusal	4.48	7.9
10	Marital disharmony	6.28	9.0
11	Change in school / medium / subjects	6.98	6.9
12	Aggressive (verbal) temperament	0.79	1.8

The standardized discriminant weights of the symptoms were obtained by dividing the vector of weights by the Mahalanobis D (5.41). The mean scores on this discriminant function was 3.1 for conduct disorder children, and it was -2.3 for hyperactivity children, and hence the cut of score was 0.4, which was the halfway mark between these two values. The procedure for application of the model along with the underlying assumptions/ instructions are as presented in the following table.

TABLE 17.3   A model for differentiating conduct disorders and hyperactivity in children: scoring key

Conduct Disorder Items:	
1. Tempted tantrum	2.1
2. Change in school / medium / subjects	1.3
3. Marital disharmony	1.2
4. Lying / stealing / cheating	0.9
5. School referred	0.8
6. Aggressive (vertical) temperament	0.2
A. Conduct Disorder Score	
Hyperactivity items:	
1. Poor attention	1.4
2. Family history of mental retardation	1.4
3. Post natal problem of child	1.0
4. Impulsive	1.0
5. Problem in language expression	0.3
6. Restless	0.1
B. Hyperactivity Score	
(A-B) : Test score	
Test diagnosis: clinical diagnosis	

Instructions:

1. If the test is more than 0.4, the child has to be classified as conduct disorder; otherwise the child is hyperactive.
2. The model is applicable only when the child to be classified as belonging to one or the other syndrome.

KEYWORDS

- **Cut-off score**
- **Hyperkinetic children**

CHAPTER 18

FACTOR ANALYSIS

CONTENTS

18.1   BASIC ELEMENTS OF FACTOR ANALYSIS

18.1.1   FACTOR LOADINGS

It is a matrix representing the correlations between different combinations of variables and factors. Let the L_{ij} is the factor loading of the ith factor on the jth variable ($i = 1, 2…k$ and $j = 1, 2 …k$). Since the factors happen to be linear combinations of variables, the coordinates of each variable is measured to obtain what are called factor loadings. Such factor loadings represent the correlation between the particular variable and the factor. The higher the factor loadings, the more likely it is that the factor is underlying that variable. Factor loadings help in identifying which variables are associated with the particular factor. There are a number of ways by which factor analysis is done.

18.1.2   EXTRACTION OF FACTORS AND ROTATION

In the process of extraction of factor loadings, the rotation helps in arriving at a suitable pattern of factor loadings by maximizing high correlation and minimizing low ones. A simple structure can be observed by looking at the factor loading of each factor. If the factor loading is high (as high as 1) on one factor and a very low factor loading (as low as 0) on the other, it is said to possess a simple structure. If there is no simple structure, then the n-dimensional space of the factors should be rotated by an angle such that the factor loadings are revised to have a simple structure that will ease the process of interpreting the factors. Rotation could be orthogonal or oblique. The orthogonal rotation should be used under the assumption that the underlying factors are uncorrelated with each other. The "varimax" method is the commonly used orthogonal rotation technique.

18.1.3   EIGENVALUES OF FACTORS

It is the sum of squares of the factor loadings of all the variables on a factor. The eigenvalue of the ith factor is given by

$$EV_i = \Sigma \ L_{ij}^2$$

Thus, the eigenvalues are the measure of amount of total variance in the data explained by a factor. Factor analysis initially considers the number of factors to be same as the total number of variables. Looking at the eigenvalue, one can determine if the factor explains sufficient amount of variance to be considered as a meaningful factor.

18.1.3.1   SCREE PLOT

An eigen value of less than 1 essentially means that the factor explains less variance than a single variable, and therefore should not be considered to be a meaningful factor. The scree plot is a graphical presentation of eigenvalues of all of the factors initially considered for extraction. It is used to decide on the number of factors to be considered.

18.1.4   COMMUNALITIES OF VARIABLES

It is the sum of squares of the factor loadings of the jth variable on all factors. Thus, it is given by

$$\text{COM}_j = \sum L_{ij}^2$$

The communalities give the variance accounted for a particular variable by all the factors. Mathematically, it is the sum of squared loadings for a variable across all the factors. The higher the value of the communality for a particular variable after extraction, the higher the amount of variance explained by factors extracted.

18.1.5   FACTOR SCORES OF PERSONS

The factor scores are obtained for each person. With factor scores, one can also perform several other multivariate statistical methods such as cluster analysis, discriminate analysis, and multi-dimensional scaling.

18.2 PRINCIPAL COMPONENTS ANALYSIS

The principal components analysis (PCA) is a variant of factor analysis; both are data analysis techniques. In PCA, all the variance in the observed variable is analyzed; whereas in other factor analysis, only shared variance is analyzed. The PCA extracts the maximum sum of squares of the loadings for each factor in turn. Accordingly, the PCA explains more variance than the loadings obtained from any other method of factoring. It is assumed that all the variables are standardized. The aim of this method is to construct new variables (P_i) called principal components that are linear combinations of a given set of variables $x_j (j = 1 \ldots k)$.

$$P_1 = L_{11}x_1 + L_{12}x_2 + \ldots + L_{1k}x_k$$

$$P_2 = L_{21}x_1 + L_{22}x_2 + \ldots + L_{2k}x_k$$

$$P_k = L_{k1}x_1 + L_{k2}x_2 + \ldots + L_{kk}x_k$$

The L_{ij} are called loadings and are worked out in such a way that the extracted principal components satisfy two conditions. First, the principal components are uncorrelated (orthogonal). Second, the first principal component has the maximum variance, the second principal component has the next maximum, and so on.

18.2.1 NUMERICAL DEMONSTRATION

The centroid method of extraction of factors is a popular technique for factor analysis because of its ease in understanding and relatively simpler calculations in comparison with other techniques. The steps involved in extracting factors using this method are explained with an example. A researcher is interested in obtaining the set of disciplines that a student exhibit while taking up a test. The scores obtained in mathematics, statistics, and in psychology are as given in the following table.

Student	Mathematics (x_1)	Statistics (x_2)	Psychology (x_3)
1	6	6	2
2	4	4	1
3	4	1	2
4	1	2	3

Student	Mathematics (x_1)	Statistics (x_2)	Psychology (x_3)
5	4	3	3
6	4	4	3
7	3	3	3
8	7	7	2
9	5	3	2
10	5	4	1

The researcher has decided to break down the data into two factors, so that the data set that has three variables can be represented by two factors.

18.2.1.1 EXTRACTION OF FIRST FACTOR

We begin with calculating the correlation coefficients matrix as shown in Table18.1(a).

TABLE 18.1 (a) Correlation coefficient matrix (R_1)

	(x_1)	(x_2)	(x_3)
x_1	1.000	0.765	−0.482
x_2	0.765	1.000	−0.271
x_3	−0.482	−0.271	1.000

Next, we check that if the correlation matrix is positive manifold. This implies that all the values of the matrix are positive. If they are not, then we have to reflect variables that have negative values. Reflection is done by assigning a weight of −1 to variables having negative values and +1 to variables having positive values. We start with looking at the values column wise. The reflected matrix is shown in Table 18.1(b).

TABLE 18.1 (b) Correlation coefficient matrix R'_1 with reflection on variable 3

	(x_1)	(x_2)	(x_3)	-
x_1	1.000	0.765	0.482	-
x_2	0.765	1.000	0.271	-
x_3	0.482	0.271	1.000	-
Sum (S_i)	2.247	2.036	1.753	6.036

Next, we calculate the first centroid factor. The sum (S_i) of each column of the matrix R'_1 is shown in the table.

$$\sqrt{T} = \sqrt{6.036} = 2.457$$

The calculations to determine the loading of factor 1 (L_{1j}) are summarized in Table 18.1(c).

TABLE 18.1 (c) Calculations to determine the loading values of the factor 1 (L_{1j})

Variable (j)	S_i	F'_1	F_1
1	2.247	0.915	0.915
2	2.036	0.829	0.829
3	1.753	0.714	−0.714

$$F'_1 = \frac{S}{\sqrt{T}} = \frac{S}{2.457}$$

18.2.1.2 EXTRACTION OF SECOND FACTOR

Next, we calculate the second centroid factor. This is done by calculating a cross-product matrix that consists of the product of each pair of coefficients loadings L_{1j} as shown in Table 18.2(a).

TABLE 18.2 (a)   Cross-product matrix

	0.915	0.829	-0.714
0.915	0.837	0.759	-0.653
0.829	0.759	0.687	-0.592
-0.714	-0.653	-0.592	0.510

The residual matrix R_2 is obtained by subtracting the cross-product matrix from R_1, and it is shown in Table 18.2(b).

TABLE 18.2 (b)   Residual matrix R_2

	x_1	x_2	x_3	
	0.163	0.006	0.171	
	0.006	0.313	0.321	
	0.171	0.321	0.490	
Sum (S_j)	0.340	0.640	0.982	1.962

$$\sqrt{T} = \sqrt{1.962} = 1.401$$

From this table, it is clear that no variable have to be reflected. The second centroid factor loading L_{2j} are summarized in Table 18.2(c), after multiplying with respective column weights.

TABLE 18.2 (c)   Calculations to determine loadings of factor 2 (L_{2j})

J	S_j	F'_2	F_2
1	0.340	0.243	0.243
2	0.640	0.457	0.457
3	0.982	0.701	0.701

$$F'_2 = \frac{S}{\sqrt{T}} = \frac{S}{1.401}$$

18.2.1.3   EXTRACTION OF THIRD FACTOR

Next, we calculate the third centroid factor. This is done by calculating a cross-product matrix that consists of the product of each pair of coefficients of loadings L_{2j}, as shown in Table 18.3(a).

TABLE 18.3 (a)   Cross-product matrix

	0.243	0.457	0.701
0.243	0.059	0.111	0.170
0.457	0.111	0.209	0.320
0.701	0.170	0.320	0.491

The residual matrix R_3 is obtained by subtracting the cross-product matrix from R_2, and it is shown in Table 18.3(b).

TABLE 18.3 (b)   Residual matrix R_3

	x_1	x_2	x_3
x_1	0.104	−0.105	0.001
x_2	−0.105	0.104	0.001
x_3	0.001	0.001	−0.001

Residual matrix R'_3 with reflections on variables 1, 2, and 3 is given in Table 18.3(c)

TABLE 18.3 (c)   Residual matrix R'_3 with reflections on variables 1, 2, and 3

	x_1	x_2	x_3	-
x_1	0.104	0.105	0.001	-
x_2	0.105	0.104	0.001	-
x_3	0.001	0.001	0.001	-
Sum (S_j)	0.210	0.210	0.003	0.423

$$\sqrt{} = \sqrt{0.423} = 0.650$$

The third centroid factor loadings are summarized in Table 18.3(d) after multiplying with the respective column weights.

TABLE 18.3 (d) Calculation to determine loadings of factor 3 (L_{3j})

J	S_j	F'_3	F_3
1	0.210	0.323	−0.323
2	0.210	0.323	−0.323
3	0.003	0.005	−0.005

$$F'_3 = \frac{S}{\sqrt{T}} = \frac{S}{0.650}$$

18.2.1.4 ANALYSIS OF THE RESULTS

Loadings of the three factors on three variables is given in Table 18.4

TABLE 18.4 Loadings of the three factors

Variable	Factor 1	Factor 2	Factor 3	Communality (COM$_j$)
x_1	0.915	0.243	−0.323	1.000
x_2	0.829	0.457	−0.323	1.000
x_3	−0.714	0.701	−0.005	1.000
Eigenvalues (EV$_i$)	2.034	0.759	0.209	3.002
Proportion of total variance	0.678	0.253	0.069	1.000
Cumulative proportion of common variance	0.678	0.931	1.000	-

Next, we assign each variable to the factor with which it has the maximum absolute loadings as shown in table. The rule for assignment of a variable to a particular factor is simple. We assign that variable to a factor that has its factor loading greater than 0.6. Therefore, looking at Table 18.4, we can say that variables and load at factor1 as their individual loadings are greater than 0.6. The final assignment of variables to factors is summarized in Table 18.5.

TABLE 18.5　Assignment of variables to factors

Factor Number	Name of the Factor	Variables	Description
1	Mathematics discipline	x_1	Mathematics, statistics
		x_2	
2	Applied discipline	x_3	Psychology
3	-	-	-

The variables of mathematics and statistics can be replaced by a suitable factor that can be termed as mathematical discipline factor. Psychology subject form a factor called applied factor. About 68 percent of the total variance explained is due to factor 1, followed by factor 2 that explains another 25 percent. Cumulatively, factor 1 and 2 explains 93 percent of the total variance, and hence we could have considered these two factors and drop the third factor.

KEYWORDS

- **Centroid factor**
- **Communality**
- **Correlation coefficients**
- **Eigenvalue**
- **Factor analysis**
- **Varimax**

CHAPTER 19

META-ANALYSIS

CONTENTS

19.1 LOCATION AND SELECTION OF STUDIES FOR META-ANALYSIS

19.1.1 LOCATION OF STUDIES

19.1.1.1 PUBLICATION BIAS

The source of search for literature in meta-analysis includes the published literature, unpublished literature, uncompleted research reports, and work in progress. Reliance on only published reports leads to publication bias-the bias resulting from the tendency to publish results that are statistically significant. Hence, the meta-analysis needs to obtain information from unpublished research also. Contact with colleagues, experts in the applied field, and other informed channels can be important sources of information on published, uncompleted, and ongoing studies.

19.1.2 SELECTION OF STUDIES

Given a vast quantity of heterogeneous literature, suitable studies have to be selected for a meta-analysis. The inclusion and exclusion criteria relate to quality and to the combinability of patients and outcome.

19.2 END-POINTS OF PRIMARY STUDIES

Meta-analysis is a two-stage process. In the first stage, the end-points (summary statistics) are collected from each primary study.

19.2.1 A LIST OF END-POINTS

The important end-points are the following:
1. Uncontrolled studies: Proportion/percentage, mean response, etc.
2. Controlled studies: Risk difference, risk ratio, odds ratio, mean difference, standardized mean difference, etc.
3. Critical ratios: Z-values, χ^2-values, t-values, F-ratios, etc.
4. Effects sizes: Glass C, Hedges g, Cohen's d, etc.

19.2.1.1 COHEN'S d

It is given by

$$d = \frac{\left(\bar{x}_1 - \bar{x}_2\right)}{s}$$

where $\bar{x}_1$ is the mean of the first group

$\bar{x}_2$ is the mean of the second group

s is the pooled estimate of the standard deviations of the two groups

where s is given by

$$s = \sqrt{\frac{\left(n_1 - 1\right)S_1^2 + \left(n_2 - 1\right)S_2^2}{\left(n_1 + n_2 - 2\right)}}$$

19.2.2 TRANSFORMATION OF END-POINTS

All the studies selected for a meta-analysis may provide different end-points (data-points). In such cases, a transformation to common end-points is necessary. It is convenient to transform different statistics to the correlation coefficient 'r' before proceeding with further analysis. Below is a list of important transformations. When the end-point is

Contingency coefficient, $r = CC$

Phi-coefficient, $r = \emptyset$

Z-value, $r = \dfrac{Z}{\sqrt{n}}$

t-value, $r = \dfrac{t}{\sqrt{t^2 + df}}$

F-value, $r = \dfrac{SS_B}{SS_T}$

Cohen's d, $r = \sqrt{\dfrac{d^2}{d^2 + 4}}$

19.3   QUALITY ASSESSMENT OF SELECTED STUDIES

Methods of quality assessment provide a systematic approach to describe primary studies and explain heterogeneity. A formal approach to decide the ultimate inclusion status of a study may be undertaken by using a penal of judges/experts.

19.3.1   A LIST OF QUALITY ASSESSMENT ITEMS

The important quality assessment items include the following:
1. The report: principal author, year of publication, source/journal, etc.
2. The study: scope, population, etc.
3. The patients: age, sex, socio-economic status, etc.
4. Study design: observational/experimental, random sampling/randomization, non-response error/attrition rate, etc.
5. Treatment: dose and timings, duration, mode of delivery, etc.
6. Effect size: effect size, sample size, nature of outcome, etc.

19.3.2   INTERNAL VALIDITY AND EXTERNAL VALIDITY

The internal validity of a study is the extent to which systematic error (bias) is minimized. Such biases are the selection bias, performance bias, detection bias, and attrition bias. The external validity is the extent to which the results of the study provide a correct basis for applicability to other circumstances. Thus, the internal validity is a prerequisite for external validity.

19.4   META-ANALYSIS MASTER SHEET

The first step in meta-analysis is to prepare a master sheet (data point-stable). The first column in the master sheet consists of the list of selected studies according to their chronological order of publications.

19.4.1 STATISTICAL HETEROGENEITY AND CLINICAL HETEROGENEITY

The last column of the table consists of their respective end-points or their transformed values in order to notice the statistical heterogeneity. The information on relevant variables (quality assessment items) is entered in the master sheet in order to notice the clinical heterogeneity.

19.5 META-ANALYSIS PLOTS

In more complex situations to understand heterogeneity and its sources, several graphs and diagrams have been established to use in meta-analysis.

19.5.1 FOREST PLOT

The forest plot is a graphical display of results of individual studies on a common scale. In this plot, the summary statistics are represented on the horizontal axis, and the individual studies are indicated on the vertical axis. The study results are represented by dots and their 95 percent confidence intervals by the lines joining them. A vertical line representing the weighted arithmetic mean of the end-points is also drawn. It allows a visual examination of the degree of heterogeneity between the study results.

Example: The forest plot showing prevalence rates of schizophrenia reported in 37 studies in India is as given in Figure 19.1.

19.5.2 FUNNEL PLOT

The funnel plot is a simple scatter plot of the end-points of individual/primary studies on the horizontal axis against the precision of the estimates on the vertical axis. The end-points from small studies will scatter more widely at the bottom of the graph, with the spread narrows among larger studies. In the absence of bias, this will resemble a symmetrical inverted funnel. Publication bias may lead to asymmetry in the funnel plot.

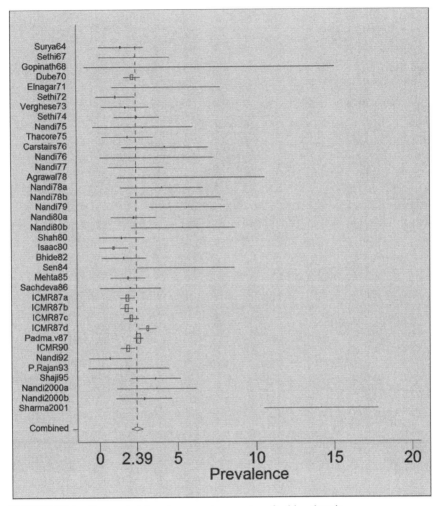

FIGURE 19.1   Forest plot showing prevalence rates of schizophrenia.

Example: The funnel plot for prevalence rates of schizophrenia reported in the 37 studies in India is as given in Figure 19.2.

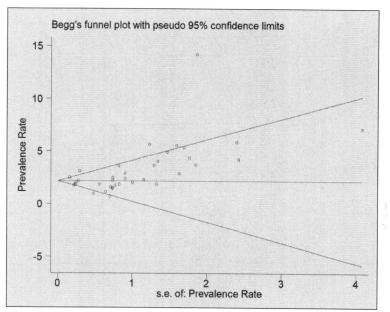

FIGURE 19.2   Funnel plot showing prevalence rates of schizophrenia.

19.6   METHODS FOR POOLING END-POINTS: FIXED EFFECTS MODEL

The methods used to pool end-points employ a weighted average of the end-points in which the larger studies generally have more influence than the smaller ones. The methods are based on the assumption such as fixed effects model and random effects model.

19.6.1   FIXED EFFECTS MODEL

The fixed effects model assumes that there is only one population, and the variability between studies as due to random variation.

19.6.2   SAMPLE-SIZE METHOD

In this method, the weights for the studies are equal to their respective sample size (weighted arithmetic mean).

19.6.2.1   POOLED ESTIMATE

The pooled estimate is given by

$$\theta_{ss} = \frac{\sum n_i \theta_i}{\sum n_i}$$

where n_i is the sample size of the ith study.

19.6.3   INVERSE VARIANCE METHOD

The inverse variance (IV) method is used to pool either the binary, continuous, or correlation data. This approach has wide applicability since it can be used to combine any estimate that has standard error available. In this method, the weights for the studies are equal to their respective precisions of the estimates.

19.6.3.1   POOLED ESTIMATE

The pooled estimate is given by

$$\theta_{IV} = \frac{\sum w_i \theta_i}{\sum w_i}$$

where the weight w_i for the ith study is given by

$$w_i = \frac{1}{SE(\theta_i)^2}$$

The standard error of the pooled estimate is given by

$$SE(\theta_{IV}) = \frac{1}{\sqrt{w_i}}$$

19.6.4   HETEROGENEITY STATISTICS

The heterogeneity statistic of selected studies is given by

$$Q_w = \sum w_i (\theta_i - \theta_{IV})^2$$

where k is the number of studies included in the meta-analysis. The Q_w follows approximate chi-square distribution with df $= (k-1)$. Any significant heterogeneity between end-points found in the selected studies should not be considered as purely a problem for meta-analysis, since it provides an opportunity for examining why treatment effects differ in different circumstances. Careful investigation of the heterogeneity in meta-analysis should increase the scientific and clinical relevance.

19.7   METHODS FOR POOLING END-POINTS: RANDOM EFFECTS MODEL

19.7.1   RANDOM EFFECTS MODEL

The random effects model assumes that there are different populations and a different underlying effect for each study, and it takes this into account as an additional source of variation. The random effects models are used when there is considerable (significant) heterogeneity between end-points.

19.7.2   DL METHOD

The DL method (DerSimonian and Laird method) of meta-analysis is the commonly used random effects model method. The effect size θ_i is assumed to have a normal distribution with mean θ and variance τ^2.

19.7.2.1   POOLED ESTIMATE

In this method, the pooled estimate is given by

$$\theta_{DL} = \frac{\sum w_i^* \theta_i}{\sum w_i^*}$$

where the weight for the ith study is given by

$$w_i^* = \frac{1}{SE\left(\theta_i\right)^2 + \tau^2}$$

The variance of the studies is given by

$$\tau^2 = \frac{Q_w - (k-1)}{\sum w_i - \frac{\sum w_i^2}{\sum w_i}}$$

This is set to zero if $Q_w \leq (k-1)$. The standard error of the pooled estimate is given by

$$SE\left(\theta_{DL}\right) = \frac{1}{\sqrt{\sum w_i^*}}$$

19.8   ADDITIONAL META-ANALYSIS TECHNIQUES

It will be advantageous to extend meta-analysis by applying the following additional meta-analysis technique in order to draw important implications.

19.8.1   SENSITIVITY META-ANALYSIS TECHNIQUE

This technique is the study of influence of different statistical methodologies on the results of meta-analysis. The methodological issues include randomization versus non-randomization, random sampling versus non-random sampling, different sample size, fixed effects model versus random effects model, and different levels of quality assessments.

19.8.2   INFLUENCE META-ANALYSIS TECHNIQUE

In this technique, the influence of each study can be estimated by deleting each in turn from the analysis, and noticing down the degree of change in the pooled estimate and its level of significance.

19.8.3   SUB-GROUP META-ANALYSIS TECHNIQUE

The sub-group meta-analysis is the study of variation by different categories of patients on the results of meta-analysis. The estimate of the overall effect is intended to guide decisions about clinical practice for a wide range of patients. When the heterogeneity statistic is significant, it is not possible to assume that the given effect is same across different groups of treatments such as males versus females, young versus elderly, and those with mild versus those with severe diseases. It seems reasonable to base treatment decision upon the results of those studies that have similar characteristics to the particular patient under consideration.

19.8.4   CUMULATIVE META-ANALYSIS TECHNIQUE

This technique is a repeated performance of meta-analysis whenever a new relevant study is available for inclusion. This allows a retrospective identification of the patient in time whenever a treatment effect first reached conventional level of statistical significance. Subsequent studies will simply confirm the original results with reduced levels of significance. Further studies in large number of patients involving high cost may be avoided by conducting timely cumulative meta-analysis.

19.9   IMPLICATIONS OF THE RESULTS OF META-ANALYSIS

19.9.1   RESEARCH IMPLICATIONS

The meta-analysis helps scientists to plan new research.

19.9.2   CLINICAL IMPLICATIONS

The meta-analysis helps psychiatrists to produce evidence-based treatment. Applying results of meta-analysis to an individual patient involve consideration of the applicability of the evidence, the feasibility of the intervention in a particular settings, the risk ratio in the individual patient, and incorporation of patients' value and preferences.

19.9.3 ECONOMIC IMPLICATIONS

The meta-analysis helps in the development of methods for economic evaluation. The major implications of meta-analysis lies in the determination of the number needed to treat (NNT). The NNT is the number of patients that must be treated over a defined time period to prevent one death or disorder. Hence, it is the reciprocal of the risk difference (RD). That is,

$$NNT = \frac{1}{RD}$$

Thus, the NNT is commonly used to summarize the beneficial effect of treatment in a clinically relevant way.

19.9.4 IMPLICATIONS FOR POLICY MAKING

The meta-analysis helps policy makers to make meaningful decisions. Policy making should be based on best current knowledge and takes into account of resources and values in the interpretation of evidence.

19.9.5 NUMERICAL DEMONSTRATION

Let suppose that we have information on proportion of schizophrenia registrations at three mental institutes (hospitals) as shown in the following meta-analysis master sheet:

Mental Institute	Ownership	Registrations	Schizophrenia	Proportion
A	private	91	8	0.0879
B	government	100	25	0.2500
C	private	109	5	0.0459
Combined	-	300	38	0.1270

We wish to obtain the pooled proportion of schizophrenia in these hospitals. As a first step, we have the weighted proportion based on sample-size method, as 0.127 or 12.7 percent. In order to apply the IV method of fixed effects model, we will have the computation table as shown below:

Institute	n	x	θ_i	$SE(\theta_i)^2$	w_i	$\Sigma w_i \theta_i$	$(\theta_i-\theta_w)^2$
A	91	8	0.0879	.00088	1136	100	0.022
B	100	25	0.2500	.00188	532	133	14.748
C	109	5	0.0459	.00040	2500	115	3.534
Combined	300	38	0.1267	-	4168	348	18.304

$$\theta_{IV} = \frac{348}{4168} = 0.0835, \text{ that is, } 8.35\%$$

$$SE(\theta_{IV}) = \frac{1}{\sqrt{4168}} = \frac{1}{64.56} = 0.0155$$

From the table, the heterogeneity statistic is given as follows:

$Q_w = 18.304$ with df $= 2$

Since the calculated value is more than the table value of $\chi^2(2) = 9.21$, the null hypothesis is rejected ($P < 0.01$, highly significant). This means the hospitals are significantly different proportions of schizophrenic cases based on ownership. Hence, we may proceed with random effects model method such as the DL method. The variance of the end-points (proportion of schizophrenic cases) of the three hospitals is calculated to be as follows:

$$\tau^2 = \frac{18.304 - (3-1)}{4168 - (\frac{7823520}{4168})} = \frac{16.304}{2291} = 0.0071$$

The computation table for DL method is as given below:

Institute	n	x	θ_i	v_i	$v_i + \tau^2$	w_i^*	$\theta_i w_i^*$
(1)	(2)	(3)	(4)	(5)	(6)	(7)	(8)
A	91	8	0.0879	.00088	.00798	125	11.00
B	100	25	0.2500	.00188	.00898	111	27.75
C	109	5	0.0459	.00040	.00750	133	6.10
Combined	300	38	0.1267	-	-	369	44.85

$$\theta_{DL} = \frac{44.85}{369} = 0.1215 \text{ or } 12.15\%$$

$$SE(\theta_{DL}) = \frac{1}{\sqrt{369}} = 0.0521$$

Thus, the pooled estimate of proportion of schizophrenia registrations of mental hospitals are 0.1270 (based on sample size method), 0.0835 (based on IV method), 0.1215 (based on DL method). Since the heterogeneity statistic is significant, we have to accept the pooled estimate based on DL method.

19.10 ESTIMATION OF PREVALENCE RATES OF MENTAL AND BEHAVIORAL DISORDERS IN INDIA

In order to obtain prevalence rates of mental and behavioral disorders in India, 13 psychiatric epidemiological studies were selected for meta-analysis, after fulfilling standard criteria.

TABLE 19.1   Prevalence rates reported in studies selected for meta-analysis

Chief Investigator		Year of Report	Number of Families	Number of Persons	Number of Cases	Prevalence Rate
1	Sethi	1967	300	1733	126	72.7
2	Elnagar	1971	184	1383	38	27.5
3	Sethi	1972	500	2691	109	39.4
4	Nandi	1975	177	1060	112	105.7
5	Nandi	1977	590	2918	170	58.3
6	Nandi	1980a	815	4053	204	50.3
7	Nandi	1980b	404	1862	242	130.0
8	Sen	1984	337	2168	99	45.7
9	Mehta	1985	1195	5941	86	14.5
10	Sachdeva	1986	376	1989	75	37.7
11	Nandi	1992	353	1424	68	47.8
12	Premarajan	1993	225	1115	106	99.4
13	Shaji	1995	1094	5284	82	15.5

Sixteen diagnostic categories were reported in these studies as listed below.

1. Organic Psychosis (OP)
2. Schizophrenia (Schi)
3. Manic Affective Psychosis (MAN)
4. Manic Depression (MD)
5. Endogenous Depression (ED)
6. Mental Retardation (MR)
7. Epilepsy (Epi)
8. Phobia (Phob)
9. Generalized Anxiety (An)
10. Neurotic Depression (ND)
11. Obsessive and Compulsive Neurosis (OCN)
12. Hysteria (Hys)
13. Alcohol /Drug Addiction(Al)
14. Somatization (Som)
15. Personality Disorders (PD)
16. Behavioral / Emotional Disorders of Children (BE)

Only four studies have covered all the 16 diagnostic categories as shown in Table 19.2. The prevalence rates per 1,000 population of each diagnostic categories of these studies were worked out as shown in the table. The sample size method of meta-analysis has been employed on the basic data. The sample size method of estimating the prevalence rate of a diagnostic category consists in omitting the studies that have not covered the diagnostic category and calculating the weighted average by taking the sample size as weights. The weighted prevalence rates of the diagnostic categories were as shown Table 19.2.

TABLE 19.2   Prevalence rates of mental and behavioral disorders reported in studies selected for meta-analysis

Chief investigator (Year)	OP	Schi	Man	MD	ED	MR	Epi	Phob	AN	ND	OCN	Hys	AL	Som	PD	BE
Sethi (1967)	-	2.3	-	-	1.7	22.5	1.2	1.2	17.9	5.2	1.2	4.0	#	0.6	1.7	13.2
Elnagar (1971)	-	4.4	-	-	2.9	1.4	4.4	-	-	-	-	1.4	13.0	#	#	#
Sethi (1972)	0.4	1.1	-	0.7	-	25.3	2.2	-	3.0	1.5	-	2.2	#	-	1.5	1.5
Nandi (1975)	-	2.8	-	-	37.7	2.8	10.4	1.0	12.3	4.7	1.0	17.0	0.9	9.4	2.8	2.8
Nandi (1977)	-	2.4	1.4	-	14.4	7.2	3.1	3.4	1.7	1.0	2.4	4.1	16.1	-	-	1.0
Nandi (1980a)	-	2.2	1.5	-	17.3	8.7	3.2	4.4	1.7	0.7	3.0	6.0	-	-	-	0.7
Nandi (1980b)	-	5.4	4.3	10.7	30.6	-	5.4	27.4	17.7	3.8	21.5	3.2	#	#	#	#
Sen (1984)	-	5.5	-	0.9	17.5	5.1	3.2	1.4	1.9	3.2	0.5	4.6	#	-	-	1.8
Mehta (1985)	0.5	1.9	-	1.5	-	3.2	7.4	#	#	#	#	#	#	#	#	#
Sachdeva (1986)	2.0	2.0	-	13.1	-	2.5	2.5	#	#	#	#	#	15.6	#	#	#
Nandii (1992)	-	0.7	2.8	-	31.6	5.6	4.2	-	1.4	-	0.7	0.7	#	#	#	#
Prema -ranjan (1993)	-	1.9	-	15.0	-	4.7	0.9	-	13.1	24.4	-	1.9	25.3	-	1.9	10.3
Shaji (1995)	0.9	3.6	-	3.0	-	2.8	5.1	#	#	#	#	#	#	#	#	#
Weighted prevalence rate	0.4	2.7	0.7	2.7	8.9	6.9	4.4	4.2	5.8	3.1	3.1	4.5	6.9	0.6	0.6	**2.7**

NIL, # Not covered

The total of all the weighted prevalence rates is 58.2. Thus, the prevalence rates of mental and behavioral disorders in India are estimated to be 58.2 per 1,000 population.

KEYWORDS

- **End-points**
- **Heterogeneity statistics**
- **Meta-analysis**

CHAPTER 20

REPORTING THE RESULTS

CONTENTS

20.1 EVALUATION OF THE STUDY

20.1.1 DATA DESCRIPTION

- The study design must be appropriate to the objectives of the study. It depends on whether intervention is involved or not.
- The target population and the sampling units/observational units must be clearly defined, and suitable sampling method must be chosen.
- The sample size must be adequate to detect the difference/relationship that we look for. That is, we should provide adequate means of precision of the estimate.
- We must have all relevant explanatory variables. The outcome variable must be clearly defined.
- The rate of missing values/non-compliance must be reasonably low. The bias that is introduced must be determined.

20.1.2 INTERNAL VALIDITY AND EXTERNAL VALIDITY

Internal validity concerns observational bias, effect of random variation, outcome variation with the magnitude of the source, strength of relationship between source of exposure and outcome, and time relationship, whereas external validity concerns the applicability of the results to target population.

20.1.3 COMPARISON WITH OTHER EVIDENCES

The consistency of the results with those of other studies as well as the plausibility of the results must be specified.

20.2 INTERPRETING THE RESULTS

In confirmatory studies, strong conclusions can be drawn about the relationships found. Otherwise, it is only exploratory and further studies become necessary.

20.2.1 DATA SET

- In experimental trials, one must consider the extent to which the subjects of some larger population are represented. However, this question does not arise in observational studies.
- Random sampling errors have decreasing importance as the sample size increases, bias does not.
- Important explanatory variables may produce outliers, which should not be missed. Confounding factors are operating that have not been unaccounted for.

20.2.2 DESCRIPTIVE STATISTICS

- The average values are not necessarily normal or typical values. The variability is also essential. In the same way, a regression line cannot be interpreted as meaning that most or all individuals would be expected to lie on it.
- The standard deviation is a direct estimate of variability in the population, whereas the standard error is a description of the precision of some estimate.

20.2.3 STATISTICAL INFERENCE

- Rejecting a null hypothesis does not mean that the alternate hypothesis is true.
- The statistical significance depends on sample size, not on scientific importance; with a large enough sample, you can detect the uninteresting minimal difference.
- In experimental studies, randomization of treatment assignments is supposed to guarantee that you can draw causal conclusions. In observational studies, we can only support causality on nonstatistical grounds. The important points that are to be considered include the strength of association between source of exposure and response; the consistency of association among different circumstances, people, place, and time; and the scientific plausibility of the explanations.
- In comparing your results with those previously obtained elsewhere, be careful about publication bias; only studies yielding statistically

significant results are generally published. One convenient method of comparing all studies on a subject is through an overview or meta-analysis.

20.3   WRITING THE REPORT

Your report will have the potentials to change what scientists or decision makers believe about the phenomenon under study. To make this possible, your report must
- be reasonably brief
- avoid long and technical words
- be well-organized and coherent
- be in the form of tables and graphs where necessary
- ensure that reading your report should be a stimulating and satisfying experience

20.3.1   MAJOR CRITICISMS

The report should modify the scientific belief by strengthening, weakening or altering existing ones, or creating new ones. For this, the design or analysis should not be flawed. Major criticism will include the following:
- Unrepresentative sample to allow generalization
- Lack of or inadequate randomization
- Key missing explanatory variables.

20.3.2   WRITING SCIENTIFIC PAPERS

The publications in scientific journals will enable your results to be rapidly disseminated throughout the world. Never publish interim results of your experimental trials if the trial is continuing. This can bias the response still to come. In a properly designed study, the so-called negative findings of no relationship among key variables can be as important as positive ones. They do not indicate failure of the study. The basic sections of scientific papers are the following:
- Title
- Abstract (summary of your main results)

- Introduction (recalling relevant previous research and justifying your objectives in the present work)
- Method (describing what you do. That is, the protocol and how you followed it)
- Results (describe what you discovered)
- Discussion (interpreting your results and drawing implications)
- References (related publications in the field)
- Appendix (containing more technical and detailed supporting material)

20.3.3 WRITING TECHNICAL REPORTS

Many studies are conducted in order to make policy decision, whether in government or in private industry. The drive to produce valid and objective results may be much stronger than in scientific publishing because concrete action will be taken. The basic sections of technical papers are the following:

- Title
- Material and methods (study design and statistical model)
- Results (analysis and presentation)
- Discussion (interpretation)

All sections, except the discussion that gives the authors opinion, are supposed to be objective. Most readers will want to master the main results, and the nature of the arguments, along with its limitations.

20.3.4 REPORTING THE DATA ANALYSIS

- Give good reasons for excluding outlying observation.
- In case of estimation of parameters, generally avoid tests in favor of intervals of precision.
- Report all the tests made, not just the significance ones; otherwise, you are committing fraud.
- A statistically significant results neither mean that it is of any practical importance, nor does it even prove that the relationship is real.

20.3.5   *REPORTING THE RESULTS*

- Give the characteristics of non-respondents and dropouts. In experimental trials, thoroughly document and discuss both noncompliance and side effects from the intervention.
- In experimental trials, describe objectively what happened to subject on each treatment. In regression line, show all the individual observational points.
- The simple descriptive statistics may serve to illustrate results from more complex analysis that the general reader cannot be expected to understand. However, clearly document those complex analyses.
- The graphs and diagrams may be more useful along with the descriptive statistics. The important graphs and diagrams are given below:

  Histogram in one-variable descriptive statistics
  Scatter plot in correlation analysis
  Regression line in regression analysis
  Survival curve in survival analysis
  Sequence chart in time series analysis
  Profiles in multivariate analysis
  Dendrogram in cluster analysis
  Scree plot in factor analysis
  Forest plot in meta-analysis

KEYWORDS

- **Descriptive statistics**
- **Experimental trials**

CHAPTER 21

STATISTICAL PACKAGE FOR SOCIAL SCIENCES

CONTENTS

21.1   SPSS DATA EDITOR

Once installed, SPSS can be opened like any other windows-based application by clicking on the start menu at the bottom left-hand corner of the screen and by clicking on SPSS for windows from the list of programs; or by clicking on the icon on the screen. This activates the SPSS data editor window as shown in the following figure.

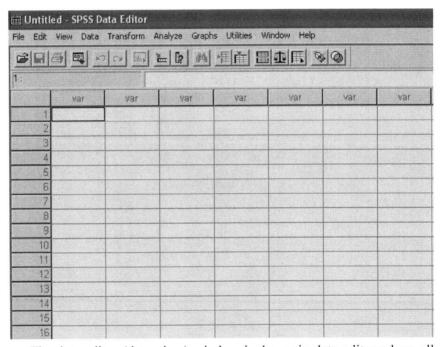

The data editor (data view) window is the main data editor where all the data are entered much like an Excel spreadsheet.

21.1.1   MENU ITEMS

At the top of the screen, there are different menus that give access to various functions of SPSS. This consists of 10 menus, which provides access to every tool of the SPSS program.

21.1.1.1   FILE, EDIT, AND VIEW

The file menu lets us open, save, print, and close file and provides access to recently used files, etc. The Edit menu allows us do things such as copy, cut, paste, etc. The View menu can hide or show the toolbar, status bar, gridlines, etc.

21.1.1.2   DATA AND TRANSFORM

The Data menu allows us to manipulate the data in various ways. We can define variables, go to a particular case, sort cases, transpose them, merge cases, as well as variables from some other file. We can also select cases on which we want to run the analysis and split the file to arrange the output of analysis in a particular manner. The Transform menu lets us compute new variables and make changes to existing ones.

21.1.1.3   ANALYZE AND GRAPH

The Analyze menu is the function that lets us perform all the statistical analysis. This has various statistical tools grouped under different categories. The Graph menu lets us make various types of plots from our data.

21.1.1.4   UTILITIES, WINDOW, AND HELP

The Utilities menu gives us information about variables and files. Finally, the Window and Help menus are very similar to other windows application menus.

21.1.2   FURTHER ITEMS

Below the menu bar, there is a toolbar that has buttons for quick access to various functions. The same functions can be performed by choosing relevant options from the menus. There are 17 items in the tool-bar—Open, Save, Print, Recall Dialogue, etc. At the bottom of the screen, we have a status bar like SPSS is ready. The program can be closed by clicking on the close button at the top right-hand side corner.

21.2 PROCEDURES FOR ENTERING DATA IN SPSS

As shown in SPSS data editor window, the bottom of the data editor has two tabs: data view and variable view. When in data view, the data editor works pretty much in the same manner as an Excel spread sheet. One can enter values in different cells, modifying them, and even cut and paste to and from an Excel spread sheet. When the data editor is in variable view, the data editor window is as shown below:

	Name	Type	Width	Decimals	Label	Values	Missing	Columns	Align	Measure
1	regn	Numeric	8	0		None	None	8	Right	Scale
2	serial	Numeric	8	0		None	None	8	Right	Scale
3	age	Numeric	8	0		None	None	8	Right	Scale
4	sex	Numeric	8	0		{1, male}..	None	8	Right	Scale
5	maritals	Numeric	8	0		{1, single}..	None	8	Right	Scale
6	locality	Numeric	8	0		{1, rural}..	None	8	Right	Scale
7	icd10	Numeric	8	1		None	None	8	Right	Scale
8										
9										
10										

In addition to entering the values of the variables, we have to provide information about them in SPSS. This can be done when the data editor is in variable view. Notice that there are 10 columns in the data editor (variable view) window. We will explain the use of each of these 10 columns in entering the data with the help of the data of 40 psychiatric patients as shown in Appendix I.

21.2.1 NAME

In our example, we have seven variables to enter. We will name registration number as "regn," serial number of the patients as "serial," age and sex can be named as it is, marital status as "marital," and so on.

21.2.2   TYPE, WIDTH, AND DECIMAL

The second column titled "type" lets us define the variable type. If we click on the cell next to variable name and in the type column, a dialogue box appears as shown in figure below.

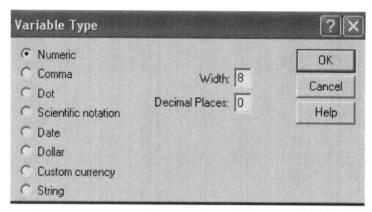

The most common type used is numeric, which means that the variable has a numeric value. The other common choice is string, which means that the available is in test format. We cannot perform any statistical analysis on a numeric variable if it is specified as a string variable. Since our variables are of numeric type, we select numeric from the dialogue box as shown in the figure above. We can also specify the width of the variable column and decimal places on this dialogue box.

21.2.3   VALUE AND LABEL

We have a column titled "label." Here, we can write the details about a particular variable in this column. We have a column labeled "values." If we click on the cell next to the variable name and in the values columns, a dialogue box appears as shown in the figure below.

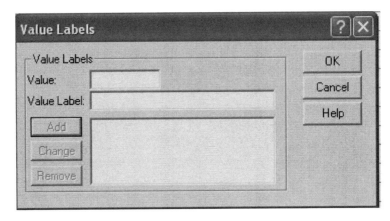

In this box, we can specify values for our qualitative variables (categories). In our data, for example, we have two values for sex: 1 representing male and 2 representing female. Enter 1 in the empty box labeled "value" and specify its name (male) in the next box labeled "value." This will activate the Add button. Click on this button and repeat these steps to specify female.

21.2.4   FURTHER ITEMS

We have a column labeled "missing" to specify missing values. By default, no missing value is selected here. The next column titled "column" helps us modify the way we wish to view the data on screen. In this column, the width of the column can be specified. In the Align column, we can specify if we want our data to be right, left, or center aligned. Finally, in the column titled "measure," we can specify whether our variable is nominal, ordinal, or scale (interval or ratio).

21.2.5   SAVING THE FILE

Once the variables are specified, we can switch to data view and enter the data. This data file can be saved just as MS Word file and reopened by double clicking on the file from its saved location. The data in Appendix I are as shown below in the data editor:

data - SPSS Data Editor

File   Edit   View   Data   Transform   Analyze   Graphs   Utilities   Window   Help

1 : regn   293307

	regn	serial	age	sex	maritals	locality	icd10	var
1	293307	1	60	1	2	1	32.3	
2	293308	2	80	2	1	1	.1	
3	293309	3	4	2	1	3	80.0	
4	293310	4	22	2	1	1	34.0	
5	293311	5	26	1	2	1	23.0	
6	293312	6	21	1	1	1	20.3	
7	293313	7	23	2	2	1	32.3	
8	293314	8	25	1	1	1	20.3	
9	293315	9	29	1	2	3	42.0	
10	293316	10	9	2	1	3	81.3	
11	293317	11	51	1	2	1	44.5	
12	293318	12	31	2	2	2	34.1	
13	293319	13	48	2	2	1	41.1	
14	293320	14	25	2	1	1	32.3	
15	293321	15	65	1	2	2	29.0	
16	293322	16	12	1	1	1	81.3	
17	293323	17	16	2	1	1	70.0	
18	293324	18	13	1	1	3	70.0	
19	293325	19	29	1	2	3	31.1	
20	293326	20	76	1	2	1	31.4	
21	293327	21	12	1	1	2	90.0	
22	293328	22	10	1	1	1	81.3	
23	293329	23	35	1	2	2	10.3	
24	293330	24	40	1	2	2	20.0	
25	293331	25	11	1	1	3	44.9	
26	293332	26	70	1	2	3	31.0	
27	293333	27	24	2	1	3	43.0	
28	293334	28	58	2	2	2	.0	
29	293335	29	60	2	2	2	23.0	
30	293336	30	27	1	1	2	30.1	
31	293337	31	30	1	1	2	43.2	
32	293338	32	50	2	1	1	29.0	
33	293339	33	26	1	1	2	31.0	
34	293340	34	25	1	1	2	29.0	
35	293341	35	27	1	1	3	42.0	
36	293342	36	19	1	1	3	31.8	
37	293343	37	33	2	2	3	33.0	
38	293344	38	35	1	1	1	34.1	
39	293345	39	24	2	2	2	31.6	
40	293346	40	44	2	2	2	33.1	

21.3   PROCEDURES TO RUN DATA ANALYSIS USING SPSS

21.3.1   DATA SET AND DATA EDITOR FORMAT

The data set to be analyzed will be with the researcher in his own format. Usually, the format will be that the subjects (patients) are represented by the rows and the variables will be represented by the columns. The data editor formats are different for different statistical method. Let us suppose that we have the data as shown in Appendix I.

21.3.2   PLAN OF ANALYSIS

The researcher has to note down as to how many cases, how many variables, and the level of measurement of each variable. The researcher must have a plan of analysis (statistical methods to be applied) and an idea of the appropriate dialogue box in the particular version of SPSS. In order to select appropriate dialogue box, click on Analyze, which will produce a dropdown menu. The Analyze menu includes descriptive statistics, compare means, etc. Let us wish to obtain frequency distribution of diagnosis (ICD-10) variable for our data. The data are entered in the data editor with an appropriate data editor format as shown above.

21.3.3   MAIN DIALOGUE BOX

Since we have planned to obtain frequencies of diagnosis variable, select *descriptive statistics* and then click on *Frequencies*, which will produce the dialogue box as shown in the figure below.

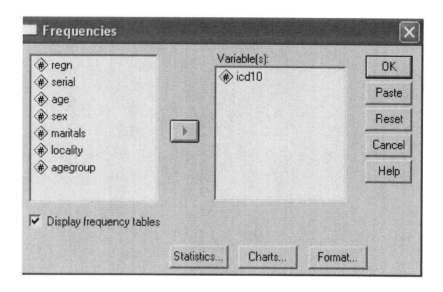

21.3.4 OUTPUT

In this dialogue box, we have to select the variables for which we wish to obtain frequencies, by transferring them to the variables box on the right-hand side from the left-hand side box. For this, first highlight the variable: ICD-10 by clicking on it. Then, click on the arrow button between the two boxes. The highlighted variable will get transferred to the other box. There are some default selections made in this box about the statistics to be shown in the output. A particular statistic can be selected or deselected for display in the output by clicking on the box next to their names. We have not selected any additional statistics other than what is selected by default. Click on OK to run the analysis.

Whenever we run any command in SPSS, the output is shown in the SPSS viewer, which opens as a separate window. The SPSS viewer window has two panels. The right-hand side panel shows the active output and the left-hand side panel shows an outline of the output shown in the right-hand side panel. One can quickly navigate through the output by selecting the same from the outline provided in the left-hand side panel. The requisite output for our data is as given below:

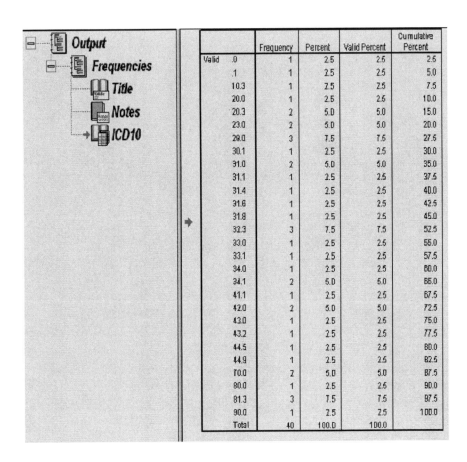

Output			Frequency	Percent	Valid Percent	Cumulative Percent
Frequencies	Valid	.0	1	2.5	2.5	2.5
		.1	1	2.5	2.5	5.0
Title		10.3	1	2.5	2.5	7.5
		20.0	1	2.5	2.5	10.0
Notes		20.3	2	5.0	5.0	15.0
		23.0	2	5.0	5.0	20.0
ICD10		29.0	3	7.5	7.5	27.5
		30.1	1	2.5	2.5	30.0
		31.0	2	5.0	5.0	35.0
		31.1	1	2.5	2.5	37.5
		31.4	1	2.5	2.5	40.0
		31.6	1	2.5	2.5	42.5
		31.8	1	2.5	2.5	45.0
		32.3	3	7.5	7.5	52.5
		33.0	1	2.5	2.5	55.0
		33.1	1	2.5	2.5	57.5
		34.0	1	2.5	2.5	60.0
		34.1	2	5.0	5.0	65.0
		41.1	1	2.5	2.5	67.5
		42.0	2	5.0	5.0	72.5
		43.0	1	2.5	2.5	75.0
		43.2	1	2.5	2.5	77.5
		44.5	1	2.5	2.5	80.0
		44.9	1	2.5	2.5	82.5
		70.0	2	5.0	5.0	87.5
		80.0	1	2.5	2.5	90.0
		81.3	3	7.5	7.5	97.5
		90.0	1	2.5	2.5	100.0
		Total	40	100.0	100.0	

21.4   DATA HANDLING USING DATA MENU

The commands described in the data menu are very useful for managing large and complex data files. The commands in data menu include select cases, weight cases, etc.

21.4.1   SPLIT FILE

Researchers may need to run analysis on a subject of the cases from the existing variables based on certain conditions. The split file commands temporarily split the file into groups. All the analysis done after splitting a file will be done separately on different subgroups, and the output will be arranged according to these groups. For example, in our data in the data editor format, we wish to obtain the frequencies of males and females separately; we have to split the file by this variable. Click on Data Menu in the data editor window and select *Split File* from the dropdown menu.

21.4.1.1 MAIN DIALOGUE BOX

The split file commands are as shown in the figure below:

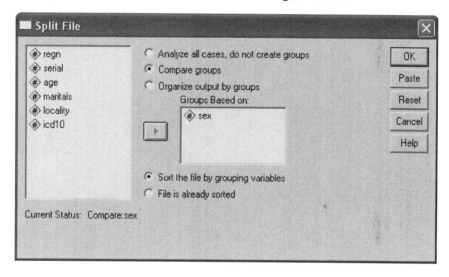

The default option in the window is to analyze all cases, do not create groups. The other options are select compare groups or organize output by groups. Selecting either of these will activate the box labeled groups based on. Select sex and transfer to this box. Click on OK to complete splitting the file. Now, if you run any analysis, two outputs will be produced, one for each of the categories. We wish to obtain the frequency distribution of diagnosis (ICD-10) according to sex. The result is as given below:

SEX	ICD-10	Frequency	Percent	Valid Percent	Cumulative Percent
Male	10.3	1	4.2	4.2	4.2
	20.0	1	4.2	4.2	8.3
	20.3	2	8.3	8.3	16.7
	23.0	1	4.2	4.2	20.8

SEX	ICD-10	Frequency	Percent	Valid Percent	Cumulative Percent
	29.0	2	8.3	8.3	29.2
	30.1	1	4.2	4.2	33.3
	31.0	2	8.3	8.3	41.7
	31.1	1	4.2	4.2	45.8
	31.4	1	4.2	4.2	50.0
	31.8	1	4.2	4.2	54.2
	32.3	1	4.2	4.2	58.3
	34.1	1	4.2	4.2	62.5
	42.0	2	8.3	8.3	70.8
	43.2	1	4.2	4.2	75.0
	44.5	1	4.2	4.2	79.2
	44.9	1	4.2	4.2	83.3
	70.0	1	4.2	4.2	87.5
	81.3	2	8.3	8.3	95.8
	90.0	1	4.2	4.2	100.0
	Total	24	100.0	100.0	
Female	.0	1	6.3	6.3	6.3
	.1	1	6.3	6.3	12.5
	23.0	1	6.3	6.3	18.8
	29.0	1	6.3	6.3	25.0
	31.6	1	6.3	6.3	31.3
	32.3	2	12.5	12.5	43.8
	33.0	1	6.3	6.3	50.0
	33.1	1	6.3	6.3	56.3
	34.0	1	6.3	6.3	62.5
	34.1	1	6.3	6.3	68.8
	41.1	1	6.3	6.3	75.0
	43.0	1	6.3	6.3	81.3
	70.0	1	6.3	6.3	87.5
	80.0	1	6.3	6.3	93.8
	81.3	1	6.3	6.3	100.0
	Total	16	100.0	100.0	-

Spitting file is a temporary option and you can cancel it by selecting analyze all cases, do not create groups.

21.4.2 WEIGHT CASES

Sometimes, the researcher may want to carry out, for example, chi-square test of significance from the frequencies in the contingency table. In such cases, the "weight cases" command in the data menu has to be used. Let us demonstrate this by a 2 × 2 contingency table as shown below:

Sex	Single	Married	Total
Male	14	10	24
Female	7	9	16
Total	21	19	40

The frequencies in 2 × 2 contingency table are entered as shown in the following data editor, where the rows (sex) and columns (marital status) are specified.

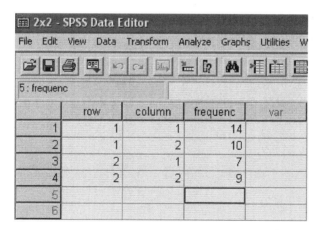

21.4.2.1 MAIN DIALOGUE BOX

Now, click on OK, which will produce a dropdown menu, choose weight cases from that, and click on the OK button.

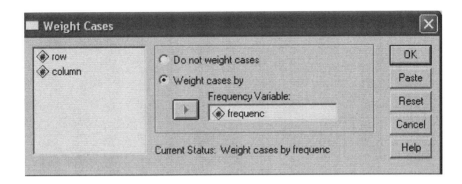

Now, click on Analyze, choose *descriptive statistics* and click on crosstab. The rows and columns are entered in their respective boxes, and now click on the OK button to run the analysis. The requisite output extracted is as shown below:

Chi-Square Tests

	Value	df	Asymp. Sig. (2-sided)	Exact Sig. (2-sided)	Exact Sig. (1-sided)
Pearson Chi-Square	.819[b]	1	.366		
Continuity Correction[a]	.338	1	.561		
Likelihood Ratio	.820	1	.365		
Fisher's Exact Test				.520	.281
Linear-by-Linear Association	.798	1	.372		
N of Valid Cases	40				

a. Computed only for a 2x2 table

b. 0 cells (.0%) have expected count less than 5. The minimum expected count is 7.60.

21.5 DATA HANDLING USING TRANSFORM MENU

21.5.1 RECODE

Sometimes, we may want to recode some of the variables. For example, we may want to create categories of age with 10 years' interval. Click on *Transform* in the data editor window and select Recode from the drop-down menu as shown in the figure below.

	regn					lity	icd10	var
1	293307					1	32.3	
2	293308			2	1	1	.1	
3	293309			2	1	3	80.0	
4	293310			2	1	1	34.0	
5	293311			1	2	1	23.0	
6	293312			1	1	1	20.3	
7	293313	7	23	2	2	1	32.3	
8	293314	8	25	1	1	1	20.3	
9	293315	9	29	1	2	3	42.0	

Menu items shown: Compute..., Random Number Seed..., Count..., Recode (Into Same Variables..., Into Different Variables...), Categorize Variables..., Rank Cases..., Automatic Recode..., Create Time Series..., Replace Missing Values..., Run Pending Transforms

Click on *"Into Different Variables"* that produces a dialogue box. Select age and transfer it into the box labeled input–output variable. Once this is done, two boxes on the left-hand side become active. We have to specify the name and label of the new variable in these boxes. Type age in that box labeled "name" and "age group" as its label in the box labeled "label." Now click on the button labeled "change" to transfer the name of the output variable into the box labeled "input variable–output variable."

21.5.1.1   MAIN DIALOGUE BOX

The resulting dialogue box appears as shown in the figure below:

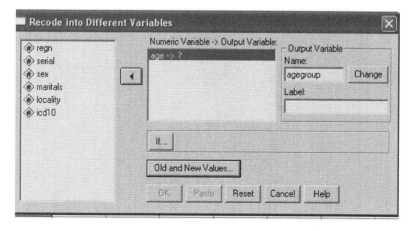

Next, click on the boxes labeled "Old and New Values," which produces a dialogue box as shown in the figure below. In this box, we have to specify our recoding strategy, which is to recode age. Specify the first range, zero through 9, specify zero in the box "New Value." Click on Add to transfer these values to the box labeled Old-->New. In a similar manner, specify the second range in the box range 10 through 19, specify 1 in the box New Value. Click on Add to transfer these values to the box labeled Old-->New, and so on. The resulting dialogue box is as shown in the figure below.

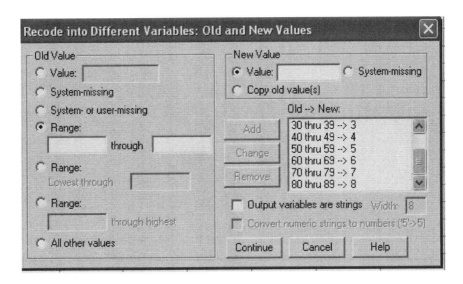

Click on "Continue" to return to the main dialogue box. Click on the OK button to complete the recoding exercise. A new variable named "age group" is created in the data file, which takes values 0, 1, 2, 3, etc., for age groups. After running on the new data editor for frequencies, the output is as shown below:

AGEGROUP

		Frequency	Percent	Valid Percent	Cumulative Percent
Valid	.00	2	5.0	5.0	5.0
	1.00	7	17.5	17.5	22.5
	2.00	14	35.0	35.0	57.5
	3.00	5	12.5	12.5	70.0
	4.00	3	7.5	7.5	77.5
	5.00	3	7.5	7.5	85.0
	6.00	3	7.5	7.5	92.5
	7.00	2	5.0	5.0	97.5
	8.00	1	2.5	2.5	100.0
	Total	40	100.0	100.0	

KEYWORDS

- **Data editor**
- **Data-handing**
- **Descriptive statistics**
- **ICD-10**
- **SPSS**

CHAPTER 22

RUNNING DATA ANALYSIS USING SPSS

CONTENTS

22.1   RUNNING "PARAMETRIC TESTS OF SIGNIFICANCE" (METHODS IN CHAPTER 9)

22.1.1   RUNNING "ONE-SAMPLE t-TEST"

Let us suppose that the improvement scores of a group of five patients are as given in the following data editor:

	patient	score	var	var	var	var
1	1	7				
2	2	4				
3	3	1				
4	4	3				
5	5	5				
6						
7						

We wish to test whether the mean score of all the patients can be assumed as 5.0.

22.1.1.1   MAIN DIALOGUE BOX

Click on *analyze* menu, which will produce dropdown menu. Choose *compare means* from that and click on *one-sample t-test*. The resultant dialogue box is as shown below. In the dialogue box, the variable (score) is selected for analysis by transferring them to the test variable box on the right-hand side. Next, change the value in the test value box, which originally appears as zero to the one against which we are testing the sample mean. In this case, the value would be 5.0.

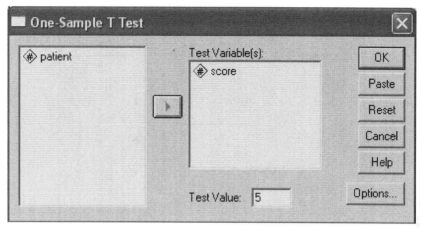

Click on the OK button to run the analysis. The output produced is as shown in the following table.

One-Sample Statistics

	N	Mean	Std. Deviation	Std. Error Mean
SCORE	5	4.00	2.236	1.000

One-Sample Test

	Test Value = 5					
					95% Confidence Interval of the Difference	
	t	df	Sig. (2-tailed)	Mean Difference	Lower	Upper
SCORE	-1.000	4	.374	-1.00	-3.78	1.78

22.1.2 RUNNING "INDEPENDENT SAMPLES t-TEST"

Let us suppose that the improvement scores of experimental patients group and a control group patient are as given in the following data editor. In the data editor, we have codes for the two groups: 1 representing the control group and 2 representing the experimental group.

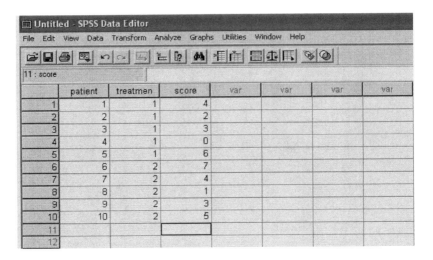

We wish to test whether the mean scores of the two groups are significantly different.

22.1.2.1   MAIN DIALOGUE BOX

Click on *analyze*, which will produce a dropdown menu, choose *compare means* from that, and click on *independent sample t-test*. The resultant dialogue box is as shown below:

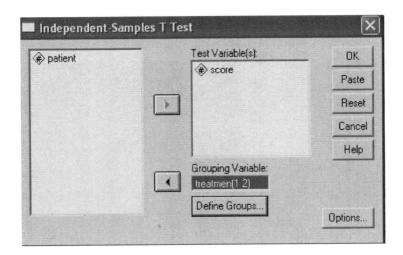

To perform independent sample *t*-test, transform dependent variable (score) into the test variable box and transfer the variable that identifies the groups (dependent) into the grouping variable box. Once the grouping variable is transferred, the "define groups" button which was earlier inactive turns active. Clicking on it will produce a box as shown below:

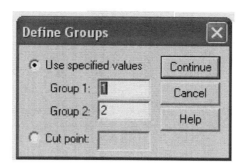

In our example, group 1 represents the control group and group 2 represents the experimental group, which we have entered under the variable group in the data editor. Therefore, put 1 in the box against group 1 and put 2 in the box against group 2, and click on "Continue." Now, click on the OK button to run the analysis. The output produced is as shown in the following table.

Independent Samples Test

		Levene's Test for Equality of Variances		t-test for Equality of Means						
		F	Sig.	t	df	Sig. (2-tailed)	Mean Difference	Std. Error Difference	95% Confidence Interval of the Difference	
									Lower	Upper
SCORE	Equal variances assumed	.000	1.000	-.707	8	.500	-1.00	1.414	-4.261	2.261
	Equal variances not assumed			-.707	8.000	.500	-1.00	1.414	-4.261	2.261

22.1.3 RUNNING "PAIRED-SAMPLE t-TEST"

The scores obtained by a group of five patients before giving the treatment, and the scores obtained by the patients after giving the treatment are given in the following data editor.

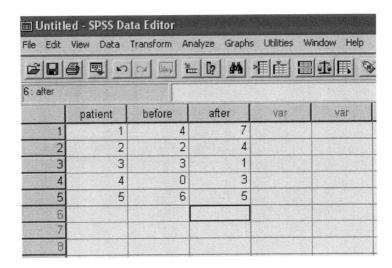

We wish to test whether the treatment program has significantly improved.

22.1.3.1   MAIN DIALOGUE BOX

Click on *analyze*, which will produce a dropdown menu, choose *compare means* from that, and click on *paired-sample t-test.* The resultant dialogue box is as shown below:

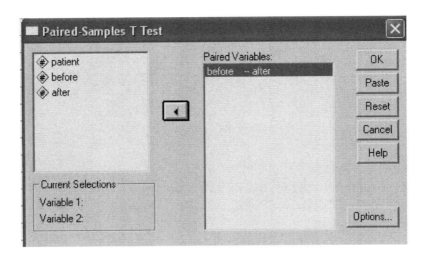

The variables to be compared (before, after) are transferred to the right-hand side box named paired variables. They will appear as before–after. Click on the OK button to run the analysis. The output produced is shown in the following fable.

Paired Samples Test

		Paired Differences							
				Std. Error	95% Confidence Interval of the Difference				
		Mean	Std. Deviation	Mean	Lower	Upper	t	df	Sig. (2-tailed)
Pair 1	BEFORE - AFTER	-1.00	2.345	1.049	-3.91	1.91	-.953	4	.394

22.2 RUNNING "ANOVA TESTS" (METHODS IN CHAPTER 10)

22.2.1 RUNNING "ONE-WAY ANOVA"

The scores obtained by 15 patients from three treatment programs using a CRD are as given in the following data editor. We have three values for treatment: 1 representing treatment 1, 2 representing treatment 2, and 3 representing treatment 3.

	patient	treat	score	var	var
1	1	1	4		
2	2	1	2		
3	3	1	3		
4	4	1	0		
5	5	1	6		
6	6	2	7		
7	7	2	4		
8	8	2	1		
9	9	2	3		
10	10	2	5		
11	11	3	10		
12	12	3	7		
13	13	3	8		
14	14	3	6		
15	15	3	4		
16					
17					
18					

We wish to test whether the mean scores of the three treatments are significantly different.

22.2.1.1   MAIN DIALOGUE BOX

Click on *analyze,* which will produce a dropdown menu, choose *compare means* from that, and click on *one-way ANOVA.* The resultant dialogue box is as shown in the following figure:

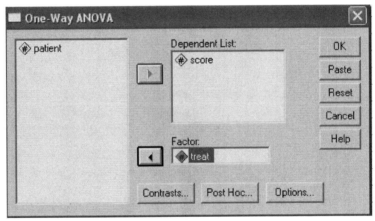

Transfer the variable score into the box labeled dependent list and the factoring variable (treatment) into the box labeled "Factor."

Click on OK to run the analysis. The output for one-way ANOVA is as shown below:

ANOVA

SCORE

	Sum of Squares	df	Mean Square	F	Sig.
Between Groups	43.333	2	21.667	4.333	.038
Within Groups	60.000	12	5.000		
Total	103.333	14			

22.2.2 RUNNING "ANCOVA TEST"

Let us suppose that the marks scored in a test by three groups of multipurpose workers before conducting a training program and the marks scored by the group after conducting the training program are as shown in the following table.

Group 1		Group 2		Group 3	
Before	After	Before	After	Before	After
4	4	6	7	8	10
2	2	5	4	6	7
1	3	3	1	9	8
0	0	3	3	8	6
3	6	8	5	4	4

The given data are entered in the data editor as shown below. The variables are labeled student, group, before, and after. For grouping variable: 1 represents group 1, 2 represents group 2, and 3 represents group 3.

We wish to test the significance of mean scores after conducting the training program and after subtracting from each individual score that portion which is predictable from the concomitant variable.

22.2.2.1   MAIN DIALOGUE BOX

Click on *analyze*, which will produce a dropdown menu, choose *general linear model* from that and click *univariate*. The resultant dialogue box is as shown below.

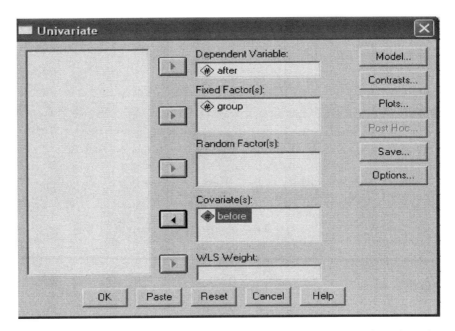

As shown in the above dialogue box, the group is transferred to the fixed factor box, the before is transferred to covariate box, and after is transferred to the dependent variable box. Now, click on the OK button to run the analysis.

The output is displayed in a table named "Tests of Between-Subjects Effects." We are interested only in the following results that are extracted from the table shown below.

Tests of Between-Subjects Effects

Dependent Variable: AFTER

Source	Type III Sum of Squares	df	Mean Square	F	Sig.
Corrected Model	76.152a	3	25.384	10.272	.002
Intercept	.721	1	.721	.292	.600
BEFORE	32.818	1	32.818	13.281	.004
GROUP	6.709	2	3.355	1.358	.297
Error	27.182	11	2.471		
Total	430.000	15			
Corrected Total	103.333	14			

a. R Squared = .737 (Adjusted R Squared = .665)

22.2.3   RUNNING "REPEATED MEASURES ANOVA"

The repeated measures scores of five students in three tests are as given in the following data editor:

We wish to test whether there is a significant improvement in these repetitions of the test by applying repeated ANOVA.

22.2.3.1   MAIN DIALOGUE BOX

Click on *analyze*, which will produce a dropdown menu, choose *general linear model* from that, and click on *repeated measures*. The resulting dialogue box is as shown below:

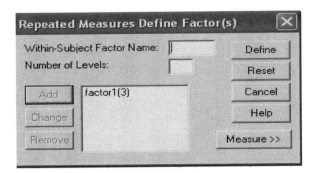

The number of levels (three in our example) is entered in the box "number of levels" as shown in the above-cited dialogue box. Then, "add" will be highlighted. Clicking on "add" will show factor (3) in the appropriate box. Click on "define" to produce the dialogue box as shown below:

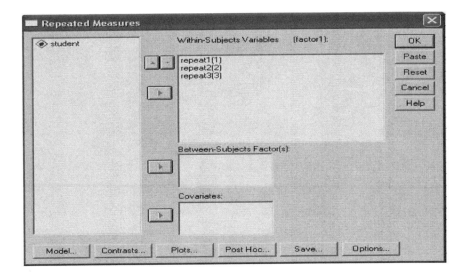

The three levels are transferred serially to the right-hand side box named Within-Subjects Variables (factor1). Click on OK to run the analysis. The SPSS display repeated ANOVA in many tables. We are interested in the following results:

Tests of Within-Subjects Effects

Measure: MEASURE_1

Source		Type III Sum of Squares	df	Mean Square	F	Sig.
FACTOR1	Sphericity Assumed	43.333	2	21.667	5.200	.036
	Greenhouse-Geisser	43.333	1.703	25.445	5.200	.046
	Huynh-Feldt	43.333	2.000	21.667	5.200	.036
	Lower-bound	43.333	1.000	43.333	5.200	.085
Error(FACTOR1)	Sphericity Assumed	33.333	8	4.167		
	Greenhouse-Geisser	33.333	6.812	4.893		
	Huynh-Feldt	33.333	8.000	4.167		
	Lower-bound	33.333	4.000	8.333		

22.3 RUNNING "NON-PARAMETRIC TESTS OF SIGNIFICANCE"(METHODS IN CHAPTER 11)

22.3.1 RUNNING "CHI-SQUARE TESTS OF SIGNIFICANCE"

The data regarding sex and marital status of the 40 registered psychiatric patients (data in Appendix I with one patient data on married is changed to single) is as presented in data editor format in Section 21.2.5. The variable sex has two categories: 1 representing male and 2 representing female. Similarly, the variable marital status has two categories: 1 representing unmarried and 2 representing married.

We wish to test whether sex and marital status is independent or not, by running chi-square test of significance.

22.3.1.1 MAIN DIALOGUE BOX

Click on *analyze*, which will produce a dropdown menu, choose *descriptive statistics* from that, and click on *crosstab*. The resulting dialogue box appears as shown below:

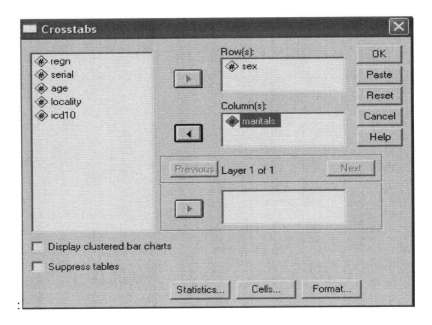

The sex is transferred to the box labeled column and marital status is transferred to the box labeled rows. Next, click on the statistics button that brings up a dialogue box. Select the first box labeled chi-square in this sub-dialogue box and click on "continue" button to return to the previous screen. Click on continue to return to the main dialogue box. Now, click on OK to run the analysis. The output produced will be as shown below:

SEX * MARITALS Crosstabulation

Count

		MARITALS		Total
		single	married	
SEX	male	14	10	24
	female	8	8	16
Total		22	18	40

Chi-Square Tests

	Value	df	Asymp. Sig. (2-sided)	Exact Sig. (2-sided)	Exact Sig. (1-sided)
Pearson Chi-Square	.269[b]	1	.604		
Continuity Correction [a]	.038	1	.846		
Likelihood Ratio	.269	1	.604		
Fisher's Exact Test				.748	.422
Linear-by-Linear Association	.263	1	.608		
N of Valid Cases	40				

a. Computed only for a 2x2 table

b. 0 cells (.0%) have expected count less than 5. The minimum expected count is 7.20.

22.3.2   RUNNING "RUN TEST"

The data regarding sex and marital status of the 40 registered psychiatric patients (data in Appendix I) are as presented in data editor format in Section 21.2.5. The variable sex has two categories: 1 representing male and 2 representing female.

We wish to test whether there is a significant randomness in the occurrence of males and females in the sequence.

22.3.2.1   MAIN DIALOGUE BOX

Click on *analyze*, which will produce a dropdown menu, choose *non-parametric test* from that, and click on *runs*. The resultant dialogue box is as given below:

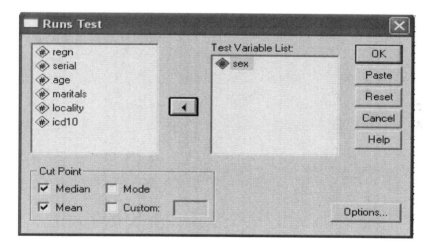

The variable sex is transferred to the box named "test variable list." The default selection for cut point is median. We have to select mean as the cut point in order to clearly distinguish between males and females. Now, click on OK to run the analysis. The requisite output is as given below:

Runs Test 2

	SEX
Test Value[a]	1.40
Cases < Test Value	24
Cases >= Test Value	16
Total Cases	40
Number of Runs	18
Z	-.568
Asymp. Sig. (2-tailed)	.570

a. Mean

22.3.3   RUNNING "MANN–WHITNEY U-TEST"

The marks scored in a test by 10 students from two departments are as given in the following data editor (1 representing the first department and 2 representing the second department).

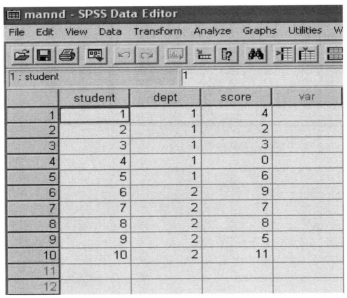

We wish to test whether the median score of the students in the two departments are significantly different by using Mann–Whitney U-test.

22.3.3.1   MAIN DIALOGUE BOX

Click on *analyze*, which will produce a dropdown menu, choose *non-parametric tests* from that, and click *two-independent sample tests*. The resultant dialogue box is as given below:

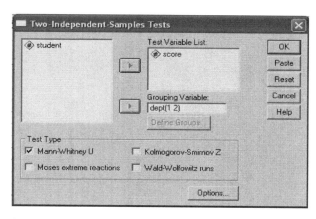

The variable score is transferred to the test variable list box. The department is transferred to the grouping variable box and defining groups by 1 and 2. Now, click on OK to run the analysis. The output is as given below:

Ranks

	DEPT	N	Mean Rank	Sum of Ranks
SCORE	1	5	3.20	16.00
	2	5	7.80	39.00
	Total	10		

Test Statistics[b]

	SCORE
Mann-Whitney U	1.000
Wilcoxon W	16.000
Z	-2.402
Asymp. Sig. (2-tailed)	.016
Exact Sig. [2*(1-tailed Sig.)]	.016[a]

a. Not corrected for ties.

b. Grouping Variable: DEPT

22.3.4   RUNNING "KRUSHKAL–WALLIS *H*-TEST"

The marks scored by 15 students in a test, from three departments are as given in the following data editor. The variables are labeled as shown in the following figure. We have three values for departments: 1 representing the first department, 2 representing the second department, and 3 representing the third department.

	student	group	score	var
1	1	1	4	
2	2	1	2	
3	3	1	3	
4	4	1	0	
5	5	1	6	
6	6	2	9	
7	7	2	7	
8	8	2	8	
9	9	2	5	
10	10	2	11	
11	11	3	14	
12	12	3	12	
13	13	3	13	
14	14	3	10	
15	15	3	16	
16				
17				

We wish to test whether the median scores of the student in three departments are same, by running the Krushkal–Wallis *H*-test.

22.3.4.1   MAIN DIALOGUE BOX

Click on *analyze* menu, which will produce a dropdown menu, choose *non-parametric tests* from that, and click on *tests for several independent-samples*. The resultant dialogue box is as shown below:

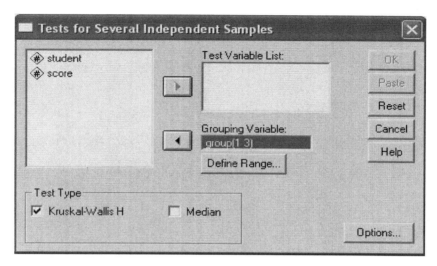

The test variables (scores) is transferred to the right-hand side box named "test variable list." The department is transferred to the grouping variable box. Define the range as 1–3 in the sub-dialogue box. Now, click on OK to run the analysis. The output is as given below:

Kruskal-Wallis Test

Ranks

	GROUP	N	Mean Rank
SCORE	1	5	3.20
	2	5	8.00
	3	5	12.80
	Total	15	

Test Statistics[a,b]

	SCORE
Chi-Square	11.520
df	2
Asymp. Sig.	.003

a. Kruskal Wallis Test

b. Grouping Variable: GROUP

22.3.5 RUNNING "FRIEDMAN TEST"

The marks scored by a group of five students in three consecutively repeated tests are as given in the following data editor. The given data are entered in the data editor and the variables are labeled as student, repeat1, repeat2, and repeat3, respectively.

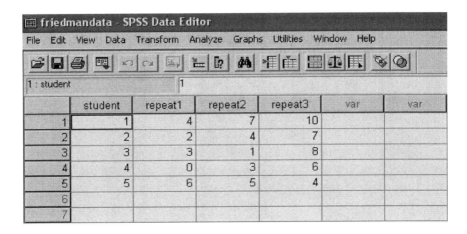

We wish to test whether there is a significant improvement in these repetitions of the test.

22.3.5.1 MAIN DIALOGUE BOX

Click on *analyze*, which will produce a dropdown menu, choose *non-parametric tests* from that, and click on *test for several independent samples*. The resulting dialogue box is as given below:

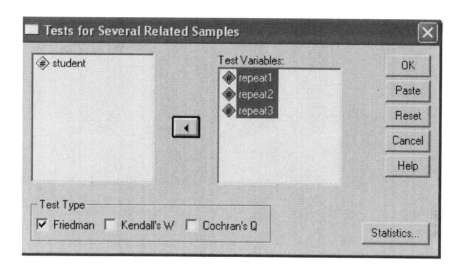

The variables repeat1, repeat2, and repeat3 are transferred to the right-hand side box named "test variable list." Now, click on OK to run the analysis. The output produced is as given below:

Ranks

	Mean Rank
REPEAT1	1.60
REPEAT2	1.80
REPEAT3	2.60

Test Statistics[a]

N	5
Chi-Square	2.800
df	2
Asymp. Sig.	.247

a. Friedman Test

22.4   RUNNING "CORRELATION ANALYSIS AND REGRESSION ANALYSIS"(METHODS IN CHAPTER 12)

22.4.1   RUNNING "PEARSON CORRELATION COEFFICIENT"

The marks scored in mathematics (mathes) and in statistics (stat) by a group of five students are as given in the following data editor.

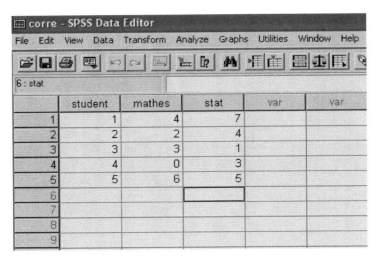

We wish to compute the Pearson correlation coefficient between the scores of mathematics and statistics using SPSS.

22.4.1.1   MAIN DIALOGUE BOX

Click on *analyze*, which will produce a dropdown menu, choose *correlate* from it, and then click on *bivariate correlation*. The resultant dialogue box is as shown below:

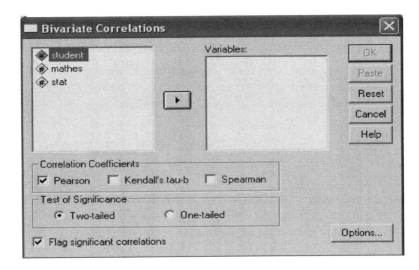

The default selection is Pearson correlation coefficient. Click on OK to run the analysis. The output produced is as shown below:

Correlations

		MATHES	STAT
MATHES	Pearson Correlation	1	.450
	Sig. (2-tailed)	.	.447
	N	5	5
STAT	Pearson Correlation	.450	1
	Sig. (2-tailed)	.447	.
	N	5	5

22.4.2 RUNNING "LINEAR REGRESSION ANALYSIS"

The "dosage" of a drug given to a group of five patients and their "improvement score" are as given in the following data editor.

	patient	dosage	improve	var	var	var
1	1	4	7			
2	2	2	4			
3	3	3	1			
4	4	0	3			
5	5	6	5			
6						
7						
8						

We wish to carry out linear regression analysis between the dependent variable (improvement) and the independent variable (dosage).

22.4.2.1   MAIN DIALOGUE BOX

Click on *analyze*, which will produce a drop down menu, choose *regression* from that, and click on *linear* as shown in the following dialogue box.

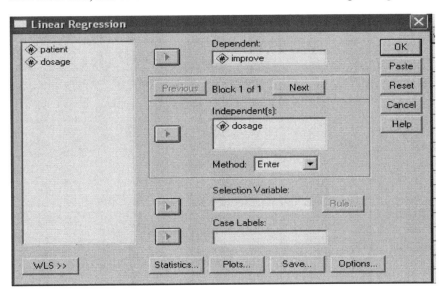

Transfer the dependent variable (improvement score) into the right-hand side box labeled "Dependent." Transfer the independent variable (dosage) into the box labeled "Independent(s)." We transfer the student into the box "Case Labels." In the main dialogue box, there are several buttons for advanced users. The "Save" button can be used to save statistics but predicted values and residuals.

Click on OK in the main dialogue box to run the analysis. The output produced has several tables. We are interested in the following output-tables.

Coefficients[a]

Model		Unstandardized Coefficients		Standardized Coefficients	t	Sig.
		B	Std. Error	Beta		
1	(Constant)	2.650	1.859		1.426	.249
	DOSAGE	.450	.516	.450	.873	.447

a. Dependent Variable: IMPROVE

22.5   RUNNING "RELIABILITY ANALYSIS AND VALIDITY ANALYSIS"(METHODS IN CHAPTER 13)

22.5.1   RUNNING "CRONBACH'S ALPHA"

The assessment scores of the services rendered by a mental institute by five raters on three items are as given in the following data editor:

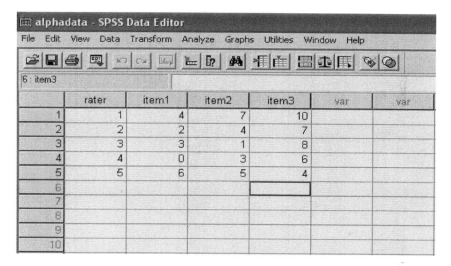

We wish to calculate Cronbach's alpha to measure whether all the three items measure the same dimension.

22.5.1.1   MAIN DIALOGUE BOX

Click on *analyze*, which will produce drop down menu, choose *scale* from that, and click on *reliability analysis*. The resultant dialogue box is as shown below:

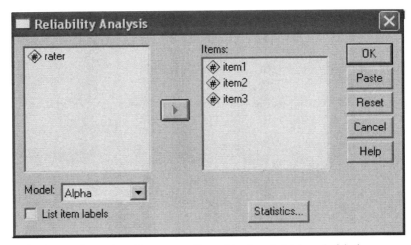

The items from 1 to 3 are transferred to the right-hand side box named "Items." Click on OK to run the analysis. The output produced is as shown below:

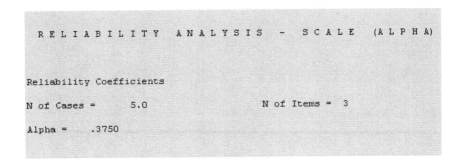

22.6   RUNNING "SURVIVAL ANALYSIS AND TIME SERIES ANALYSIS" (METHODS IN CHAPTER 14)

22.6.1   RUNNING "LIFE TABLES"

The duration of stay (DOS) of a group of 20 patients discharged in a psychiatric emergency ward are as given in the following data editor. We have only one value for status: 1 representing discharge.

	patient	dos	status	var	var	var
1	1	0	1			
2	2	1	1			
3	3	2	1			
4	4	2	1			
5	5	2	1			
6	6	2	1			
7	7	2	1			
8	8	2	1			
9	9	2	1			
10	10	3	1			
11	11	3	1			
12	12	3	1			
13	13	3	1			
14	14	3	1			
15	15	4	1			
16	16	4	1			
17	17	4	1			
18	18	5	1			
19	19	5	1			
20	20	6	1			
21						
22						
23						

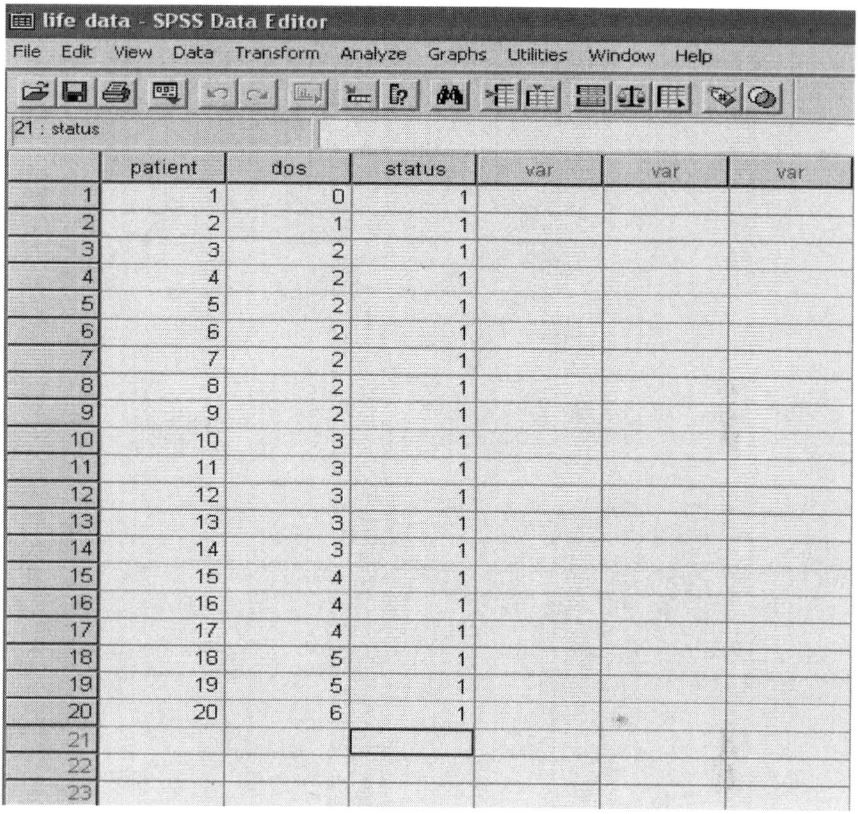

We wish to construct current hospital stay table for the data.

22.6.1.1 MAIN DIALOGUE BOX

Click on *analyze,* which will produce a drop down menu, choose *survival analysis* from that, and click on *life table*. The resultant dialogue box is as shown below:

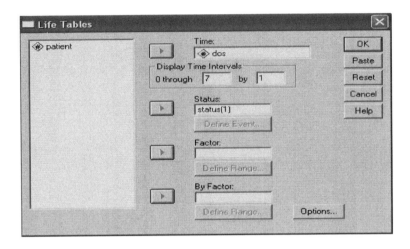

The variable (DOS) is transferred to the box named "Time." We have to specify the time interval for our data in the next down specified button. We have specified the time range as 7 (through 7) and the internal as 1 (by 1) as shown in the dialogue box. The SPSS program asks for the status of the patient after the stay in the hospital, since all the 20 patients are discharged and 1 represents the discharge. We put 1 in the status box.

Click on OK to run the analysis. The output is displayed in several tables. We are interested only in the following life table.

```
This subfile contains:      20 observations

 Life Table
    Survival Variable  DOS

        Number  Number  Number  Number                  Cumul
 Intrvl Entrng  Wdrawn  Exposd    of    Propn   Propn  Propn  Proba-
 Start   this   During    to    Termnl  Termi-  Sur-   Surv   bility  Hazard
 Time   Intrvl  Intrvl   Risk   Events  nating  viving at End  Densty  Rate
 ------ ------  ------  ------  ------  ------  ------ ------  ------  ------
    .0   20.0      .0    20.0     1.0   .0500   .9500  .9500   .0500   .0513
   1.0   19.0      .0    19.0     1.0   .0526   .9474  .9000   .0500   .0541
   2.0   18.0      .0    18.0     7.0   .3889   .6111  .5500   .3500   .4828
   3.0   11.0      .0    11.0     5.0   .4545   .5455  .3000   .2500   .5882
   4.0    6.0      .0     6.0     3.0   .5000   .5000  .1500   .1500   .6667
   5.0    3.0      .0     3.0     2.0   .6667   .3333  .0500   .1000  1.0000
   6.0    1.0      .0     1.0     1.0  1.0000   .0000  .0000   .0500  2.0000

 The median survival time for these data is   3.20
```

22.6.2   RUNNING "SEQUENCE CHART"

Let us assume that the monthly number of registrations of psychiatric disorders in a health clinic are as given in Section 14.2.2.5 is entered in the data editor. We wish to draw a sequence chart in order to configurate the time series components of the data.

22.6.2.1   MAIN DIALOGUE BOX

Click on *graph menu*, which will produce dropdown menu, select, and click *sequence*. The resultant dialogue box is as shown below:

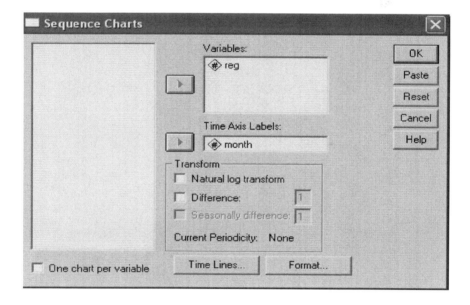

Click on OK to run the analysis. The output is displayed as follows:

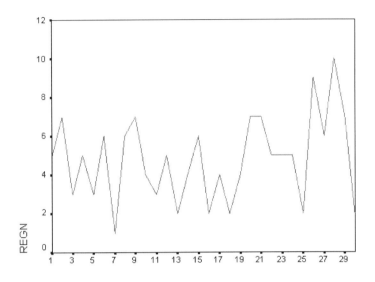

22.7   RUNNING "MULTIVARIATE STATISTICAL METHODS" (METHODS IN CHAPTER 15)

22.7.1   RUNNING "PARTIAL CORRELATION COEFFICIENT"

The scores obtained in mathematics, statistics, and psychology of a group of five students are as given in the data editor below:

	student	mathes	statisti	psychol	var
1	1	4	7	10	
2	2	2	4	7	
3	3	3	1	8	
4	4	0	3	6	
5	5	6	5	4	
6					
7					
8					

We wish to carry out partial correlation between mathes and statistics after eliminating the influence of psychology.

22.7.1.1 MAIN DIALOGUE BOX

Click on analyze, which will produce a drop down menu, choose correlation from that and click on 'partial…' resulting a dialogue box as shown below:

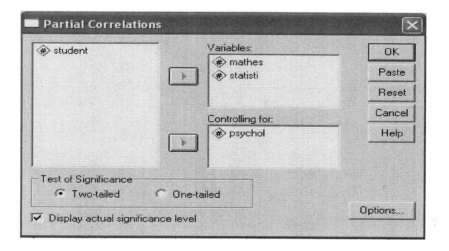

Transfer the variables mathes and statistics to the box labeled "Variables," and psychology to the box labeled "Controlling for" as shown in the dialogue box.

Click on OK in the main dialogue box to run the analysis. The output produced has several tables. We are interested in the following output table.

```
- - - P A R T I A L   C O R R E L A T I O N   C O E F F I C I E N T S

Controlling for..     PSYCHOL

              MATHES    STATISTI

MATHES        1.0000       .4955
              (    0)    (    2)
              P= .       P= .504

STATISTI       .4955     1.0000
              (    2)    (    0)
              P= .504    P= .

(Coefficient / (D.F.) / 2-tailed Significance)

" . " is printed if a coefficient cannot be computed
```

22.7.2 RUNNING "MULTIPLE REGRESSION ANALYSIS"

The IQ and dosage of a drug given to a group of five patients and their "improvement score" are as given in the following data editor:

📧 mradata - SPSS Data Editor				
File Edit View Data Transform Analyze Graphs Utilities Window Help				
6 : improvem		Redo		
	patient	iq	dosage	improvem	var
1	1	91	4	7	
2	2	100	2	4	
3	3	109	3	1	
4	4	97	0	3	
5	5	103	6	5	
6					
7					

We wish to carry out multiple regression analysis between the dependent variable (improvement) and the Independent variables (IQ and dosage).

22.7.2.1   MAIN DIALOGUE BOX

Click on *analyze*, which will produce a dropdown menu, choose *regression* from that and click on *linear...*   resulting a dialogue box as shown below.

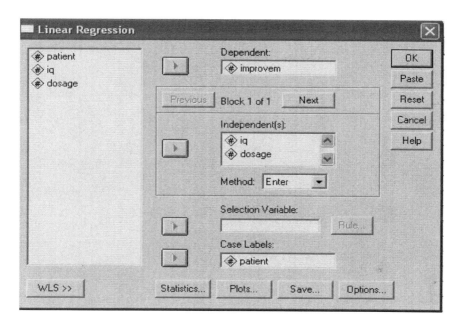

Transfer the dependent variable 'Improvement' into the right-hand side box labeled "Dependent." Transfer the independent variables (IQ and dosage) into the box labeled "Independent(s)." We transfer the patient into the box "Case Labels." In the main dialogue box, there are several buttons for advanced users. The save button can be used to save statistics such as predicted values and residuals.

Click on OK in the main dialogue box to run the analysis. The output produced has several tables. We are interested in the following output tables.

Coefficients[a]

Model		Unstandardized Coefficients		Standardized Coefficients		
		B	Std. Error	Beta	t	Sig.
1	(Constant)	31.833	3.932		8.097	.015
	IQ	-.296	.040	-.887	-7.466	.017
	DOSAGE	.583	.119	.583	4.906	.039

a. Dependent Variable: IMPROVEM

mradata - SPSS Data Editor

File   Edit   View   Data   Transform   Analyze   Graphs   Utilities   Window   Help

6 : improvem

	patient	iq	dosage	improvem	pre_1	res_1	var
1	1	91	4	7	7.246	-.246	
2	2	100	2	4	3.417	.583	
3	3	109	3	1	1.338	-.338	
4	4	97	0	3	3.138	-.138	
5	5	103	6	5	4.862	.138	
6							
7							

22.8   RUNNING "CLUSTER ANALYSIS AND DISCRIMINANT ANALYSIS" (METHODS IN CHAPTERS 16 AND 17)

22.8.1   RUNNING "HIERARCHICAL CLUSTER ANALYSIS"

The scores of 6 patients on two diagnostic tests are as given in the following data editor:

	name	test1	test2	var	var
1	A	3	6		
2	B	4	4		
3	C	5	8		
4	D	1	2		
5	E	2	0		
6	F	3	4		
7					
8					
9					

We wish to carry out cluster analysis in order to make groups of similar patients (diagnoses). We have planned to carry out hierarchical agglomerative cluster analysis using squared Euclidean distance employing nearest neighborhood method in order to get two clusters.

22.8.1.1   MAIN DIALOGUE BOX

Click on *analyze*, which will produce a dropdown menu, choose *classify* from that, and click on *hierarchical cluster*. The resultant dialogue box is as shown in the figure below:

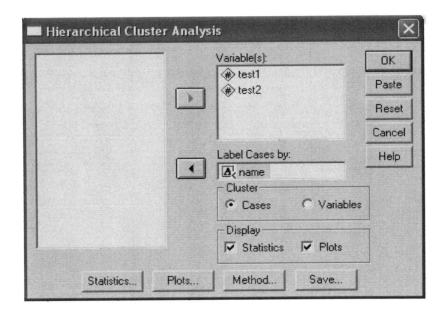

The two numerical variables (test1, test2) are transferred to the right-hand side box named "Variable(s)." The patient name labeled "Name" is transferred to the box labeled "Label Cases by." There are four buttons for advanced users. We will use the statistics button to select proximity matrix, and cluster membership as a single solution with 2 clusters. We can use "Plot" button to select dendrogram. We can use "Method" button to select neighbor method. Finally, we can use the "Save" button to save the cluster membership in the data editor.

Click on OK to run the analysis. The SPSS display cluster analysis output in many tables. We are presenting only the following output tables.

Proximity Matrix

Case	Squared Euclidean Distance					
	1:A	2:B	3:C	4:D	5:E	6:F
1:A	.000	5.000	8.000	20.000	37.000	4.000
2:B	5.000	.000	17.000	13.000	20.000	1.000
3:C	8.000	17.000	.000	52.000	73.000	20.000
4:D	20.000	13.000	52.000	.000	5.000	8.000
5:E	37.000	20.000	73.000	5.000	.000	17.000
6:F	4.000	1.000	20.000	8.000	17.000	.000

This is a dissimilarity matrix

Cluster membership and dendrogram

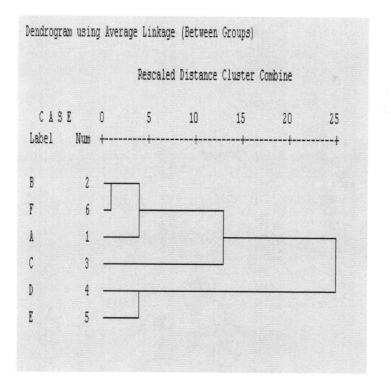

22.8.2   RUNNING "LINEAR DISCRIMINANT ANALYSIS"

The scores of two diagnostic groups of patients on two diagnostic criterion tests are as given in the following data editor:

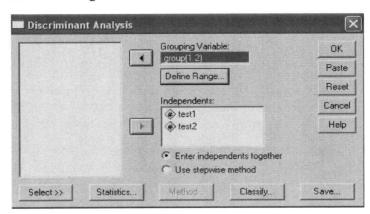

We wish to carry out discriminate analysis for the two groups.

22.8.2.1   MAIN DIALOGUE BOX

Click on *analyze*, which will produce a dropdown menu, choose *classify* from that, and click on *discriminate*. The resultant dialogue box is as shown in the following table:

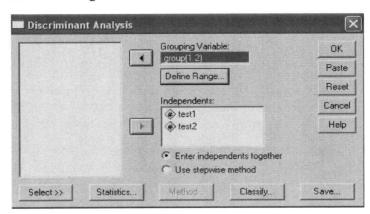

The group variable is transferred to the right-hand side box named "Grouping Variable." The two-independent variables (test 1, test 2) are transferred to the right-hand side box named "Independents." There are buttons for advanced users. We use only the "Save" button and select (1) predicted group membership, (2) discriminant score, and (3) probabilities of group membership. SPSS display discriminant analysis output in many tables. But we are interested only in the predicted group membership, scores, and probability of group membership. These results are saved in the data editor window as shown below:

Discriminant analysis - SPSS Data Editor

File   Edit   View   Data   Transform   Analyze   Graphs   Utilities   Window   Help

7 : test2

	name	group	test1	test2	dis_1	dis1_1	dis1_2	dis2_2
1	A	1	3	6	1	.57735	.79139	.20861
2	B	1	4	4	1	.57735	.79139	.20861
3	C	1	5	8	1	2.30940	.99520	.00480
4	D	2	1	2	2	-1.73205	.01799	.98201
5	E	2	2	0	2	-1.73205	.01799	.98201
6	F	2	3	4	1	.00000	.50000	.50000
7								
8								

22.9   RUNNING "FACTOR ANALYSIS" (METHODS IN CHAPTER 18)

22.9.1   RUNNING "PRINCIPAL COMPONENT ANALYSIS"

The marks scored in mathematics (mathes), statistics (stat), and in psychology (psycho) by a group of five students are as given in the following data editor:

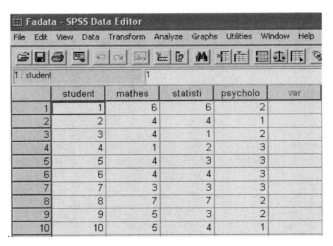

We wish to carry out factor analysis in order to reduce the 3-variables data into 2-dimensional (factors) without losing much information. We are planning to use correlation matrix and apply principal components analysis method of extraction of factors, and by applying varimax method of factor rotations.

22.9.1.1   MAIN DIALOGUE BOX

Click on *analyze,* which will produce a dropdown menu, choose data *reduction* from that, and click on *factor analysis.* The resultant dialogue box is as shown in the following dialogue box.

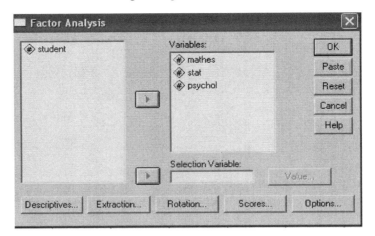

The three variables are transferred to the box named variables. In this main dialogue box, there are five buttons for advance users. We will use the descriptive button to display the correlation coefficient matrix. The extraction button is used to specify the number of factors (3 in our example). The default settings are for principal components. Finally, the "Rotation" button is used to specify method of rotation (varimax). Click on OK to run the analysis. The SPSS display factor analysis output in many tables. Only the following table is presented.

Factor Analysis

Correlation Matrix

		MATHES	STATISTI	PSYCHOLO
Correlation	MATHES	1.000	.765	-.482
	STATISTI	.765	1.000	-.271
	PSYCHOLO	-.482	-.271	1.000

Rotated Component Matrix[a]

	Component		
	1	2	3
MATHES	.463	-.274	.843
STATISTI	.925	-.105	.364
PSYCHOLO	-.104	.975	-.197

Extraction Method: Principal Component Analysis.

KEYWORDS

- **Cronbach's alpha**
- **Duration of stay**
- **One-way ANOVA**
- **t-test**

BIBLIOGRAPHY

STATISTICAL METHODS (CHAPTERS 1–20)

1. Sundar Rao, P. S. S.; Richard, J. An Introduction to Biostatistics. *A Manual for Students in Health Sciences*, 3rd ed.; Prentice-Hall of India: New Delhi, **1999**.
2. Bailley, N. T. J. *Statistical Methods in Biology*; Cambridge University Press: New York, **1994**.
3. Reddy, M. V. *Statistics for Mental Health Care Research*; NIMHANS Publications: Bangalore, **2002**.
4. Dunn, G. *Statistics in Psychiatry*; Arnold: London, **2000**.
5. Banerjee, S; Roy, R. *Fundamentals of Research Methodology*; Kitab Mahal: Allahabad, **2008**.
6. Rajagopal, V. *Selected Statistical Tests*; New Age International: New Delhi, **2006**.
7. Dawson, B; Trapp, R. G. *Basic and Clinical Biostatistics,* 4th ed.; McGraw-Hill: New York, **2004**.
8. Siegel, S. *Non-Parametric Statistics for the Behavioral Sciences*; McGraw-Hill: New York, **1988**.
9. Hope, K. *Methods of Multivariate Analysis*; University of London Press: London, **1968**.
10. Everitt, B. S. *Medical Statistics from A to Z*; Cambridge University Press: New York, **2009**.
11. Reddy, M. V.; Chandrasekhar, C. R. Prevalence of Mental and Behavioral Disorders in India: A Meta-analysis. *Indian J. Psyc.* **1998**, *40*, 149–157.
12. Bartko, J. J.; Carpenter, W. T. On the Methods and Theory of Reliability. *J. Nerv. Ment. Dis.* **1976**, *163*, 307–317.
13. Hanji, M. B. Meta-analytical Approach to Estimate Pattern of Prevalence of Schizophrenia and Epilepsy in India. Ph.D. Dissertation, NIMHANS: Bangalore, **2007**.
14. Suresh, K. P. Evaluation of Clustering Methods for Pattern Recognition in Scholastic Improvement of Rural Children. Ph.D. Dissertation, National Institute of Mental Health and Neurosciences (NIMHANS): Bangalore, **2008**.
15. Gour, A. S.; Gour, S. S. *Statistical Methods for Practice and Research*; Response Books (Sage Publications): New Delhi, **2006**.
16. Kerr, A. W.; Hall, H. K.; Kozub, S. A. *Doing Statistics with SPSS*; Sage Publications: New Delhi, **2002**.
17. Field, A. *Discovering Statistics Using SPSS*; Sage Publications: New York, **2007**.
18. Carver, R. H; Nash, J. G. *Doing Data Analysis with SPSS Version 14*; Thomson Brook Cole: Belmont, **2006**.
19. Reddy, M.V. Mental Health Delivery System by Mental Hospitals in India. *NIMHANS J.*, **1988**, *6*, 97–106.

20. Reddy, M.V. Distribution of Mental Health Manpower: An International Scene. *Indian Society for Medical Statistics Bulletin*, 1992, Vol. 11, pp 4–7.

21. Mercer, E. R. An International Psychiatric Directory. *American Psychiatric Association*; APA Press: Washington DC, 1993.

22. Anderberg, M. R. *Cluster Analysis for Applications*; Academic Press: New York, 1973.

23. Reddy, M. V. A Census of Long-stay Patients in Government Mental Hospitals in India. *Indian J. Psyc.* **2001,** *43*, 25–31.

24. Reddy, M. V. Psychiatric Patients in India: Pattern of Representation at Government Mental Hospitals and at General Hospital Psychiatric Units in India (Unpublished Paper).

25. Reddy, M. V. Government Mental Hospitals in India: Development of a Model for Optimizing Service Indicators (Unpublished paper).

26. Baker, F. B.; Hubert, L. J. Measuring the Power of Hierarchical Cluster Analysis. *J. Am. Stat. Assoc.* **1975,** *70*, 31–38.

27. Hubert, L. J.; Lavin, J. R.; A General Statistical Framework for Assessing Categorical Clustering in Free Recall. *Psychological Bulletin*, 1976, Vol. 83, pp 1072–1080.

28. Jancy, R. C. Multidimentional Group Analysis. *Aust. J. Bot.* **1966,** *14*, 127–130.

29. Johnson, S. C. Hierarchical Clustering Schemes. *Psychometrika* **1967,** *32*, 241–254.

30. MeClain, J.; Rao, B. R. CLUSTISZ: A Program to Test for the Quality of Clustering of a Set of Objects. *J. Market. Res.* **1975,** *12*, 456–460.

31. Milligan, G. W. An Examination of the Effects of Six Types of Error Perturbation of Fifteen Clustering Algorithms. *Psychometrika* **1980,** *45*, 325–342.

32. Milligan, G. W. A Monte Carlo Study of Thirty Internal Criterion Measures for Cluster Analysis. *Psychometrika* **1981,** *46*, 187–199.

33. Milligan, G. W.; Cooper, M. C. An Examination of Procedures for Determining the Number of Clusters in a Data Set. *Psychometrika* **1985,** *50*, 159–179.

34. Rohlf, F. J. Methods of Comparing Classifications. *Annu. Rev. Ecol. Syst.* **1974,** *5*, 101–113.

35. Williams, W. T.; Lance, G. N.; Dale, M. B.; Clifford, H. T. Controversy Concerning the Criteria for Taxonometric Strategies. *Comput. J.* **1971,** *14*, 162–165.

36. Verghese, A.; Abraham, A. *Introduction to Psychiatry*; The Christian Literature Society: Chennai, 1976.

37. Hanji, M. B. *Meta-analysis in Psychiatry Research: Fundamental and Advanced Methods*; Apple Academic Press: New Jersey, 2017.

APPENDIXES

APPENDIX I: BASIC DATA OF 40 REGISTERED PATIENTS AT NIMHANS

Registration Number	Serial Number	Age	Sex	Marital Status	Locality	Diagnosis (ICD-10)
293307	1	60	Male	Married	Rural	32.3
293308	2	80	Female	Single	Rural	00.1
293309	3	4	Female	Single	Urban	80.0
293310	4	22	Female	Single	Rural	34.0
293311	5	26	Male	Married	Rural	23.0
293312	6	21	Male	Single	Rural	20.3
293313	7	23	Female	Married	Rural	32.3
293314	8	25	Male	Single	Rural	20.3
293315	9	29	Male	Married	Urban	42.0
293316	10	9	Female	Single	Urban	81.3
293317	11	51	Male	Married	Rural	44.5
293318	12	31	Female	Married	Semi-urban	34.1
293319	13	48	Female	Married	Rural	41.1
293320	14	25	Female	Single	Rural	32.3
293321	15	65	Male	Married	Semi-urban	29.0
293322	16	12	Male	Single	Rural	81.3
293323	17	16	Female	Single	Rural	70.0
293324	18	13	Male	Single	Urban	70.0

Registration Number	Serial Number	Age	Sex	Marital Status	Locality	Diagnosis (ICD-10)
293325	19	29	Male	Married	Urban	31.1
293326	20	76	Male	Married	rural	31.4
293327	21	12	Male	Single	Semi-urban	90.0
293328	22	10	Male	Single	Rural	81.3
293329	23	35	Male	Married	Semi-urban	10.3
293330	24	40	Male	Married	Semi-urban	20.0
293331	25	11	Male	Single	Urban	44.9
293332	26	70	Male	Married	Urban	31.0
293333	27	24	Female	Single	Urban	43.0
293334	28	58	Female	Married	Semi-urban	00.0
293335	29	60	Female	Married	Semi-urban	23.0
293336	30	27	Male	Single	Semi-urban	30.1
293337	31	30	Male	Single	Semi-urban	43.2
293338	32	50	Female	Single	Rural	29.0
293339	33	26	Male	Single	Semi-urban	31.0
293340	34	25	Male	Single	Semi-urban	29.0
293341	35	27	Male	Single	Urban	42.0
293342	36	19	Male	Single	Urban	31.8
293343	37	33	Female	Married	Urban	33.0
293344	38	35	Male	Single	Rural	34.1
293345	39	24	Female	Married	Semi-urban	31.6
293346	40	44	Female	Married	Semi-urban	33.1

APPENDIX II: CLASSIFICATION OF STATISTICAL METHODS

CHAPTER 1: PSYCHIATRIC RESEARCH

1.1 RESEARCH QUESTIONS IN PSYCHIATRY
Recognition
Diagnosis
Treatment
Prevention

1.2 RESEARCH APPROACHES
(1) Scientific Approach
   Features
   Steps
   Science and psychiatry
(2) Mill's Canons
   Method of difference
   Method of agreement
   Joint method of difference and agreement
   Method of concomitant variation
   Method of residues
(3) Qualitative Research
(4) Quantitative Research

1.3 PROTOCOL WRITING FOR QUANTITATIVE STUDIES
(1) The Research Question
   Review of literature
   Need for the study
   Aims and objectives
(2) Type of Study
(3) Plan of Study
   Population
   Variables
   Standardization
   Experimental settings (if any)
   Record form
   Response error
   Non-response error
(4) Plan of Analysis
   Statistical methods
   Database and data analysis software
(5) Reporting the Results

1.4 VARIABLES IN PSYCHIATRY
(1) Nature of Variables in Psychiatry

(2) Qualitative Variables and Quantitative Variables
(3) Levels of Measurement of Variables
     Nominal level
     Ordinal level
     Interval level
     Ratio level
(4) Further Types of Variables
     Independent and dependent variables
     Covariates
     Concomitant variables
     Factors/dimensions
(5) Discrete and Continuous Data

1.5 STATISTICAL METHODS IN PSYCHIATRIC RESEARCH

(1) Organization and Collection of Data
     Observational studies (Chapter 2)
     Experimental studies (Chapter 3)
(2) Descriptive Statistics
     One-variable descriptive statistics (Chapter 4)
     Mental health statistics (Chapter 5)
(3) Basis of Statistical Inference
     Probability and probability distributions (Chapter 6)
     Sampling theory and methods (Chapter 7)
     Basic elements of statistical Inference (Chapter 8)
(4) Tests of Significance of Hypotheses
     Parametric tests of significance (Chapter 9)
     Experimental data analysis: ANOVA (Chapter 10)
     Non-parametric tests of significance (Chapter 11)
(5) Correlational Variables Data Analysis
     Correlational analysis and regression analysis (Chapter 12)
     Reliability analysis and validity analysis (Chapter 13)
     Survival analysis and time series analysis (Chapter 14)
(6) Multivariate Data Analysis
     Multivariate statistical methods (Chapter 15)
     Cluster analysis (Chapter 16)
     Discriminant analysis (Chapter 17)
     Factor analysis (Chapter 18)
(7) Meta-analysis (Chapter 19)

1.6 STEPS IN WRITING A PROTOCOL FOR QUANTITATIVE STUDIES

CHAPTER 2: OBSERVATIONAL STUDIES

2.1 CASE-SERIES ANALYSIS
Observational units
Variables
A proforma to obtain basic data of psychiatric patients
Record form

2.2 CROSS-SECTIONAL STUDIES
(1) Hospital In-Patients Census
Observational units
Variables
Record form
(2) Mental Morbidity Studies
Observational units
Variables
Household information schedule
Socio-economic status schedule
Case detection schedule
Case record schedule

2.3 LONGITUDINAL STUDIES
(1) Retrospective Studies
Schematic representation
Patient group and normal control group
Suspected risk/causal factors
Record form
(2) Prospective Studies
Schematic representation
With risk subjects and without risk subjects
Anticipated disease
Record form

CHAPTER 3: EXPERIMENTAL STUDIES

3.1 BASIC ELEMENTS OF EXPERIMENTS
(1) Experimental Error
(2) Basic Principles of Experiments
Replication
Randomization
Local control

CHAPTER 6: PROBABILITY AND PROBABILITY DISTRIBUTIONS

6.1 PROBABILITY SCALE
  (1) Definitions of Probability
      Theoretical probability
      Practical probability
  (2) Laws of Probability
      Addition law
      Multiplication law
      Complementary probability
      Conditional probability
  (3) Bayes, Theorem
      Formula
  (4) Evaluation of Screening Tests
      Sensitivity and specificity
  (5) Dealing with Confidential Information
      An example
6.2 PROBABILITY DISTRIBUTIONS
  (1) Binomial Distribution
  (2) Poisson Distribution
  (3) Normal Distribution
      Central limit theorem
  (4) Standard Errors of Statistics
      Standard error of sample mean
      Standard error of sample proportion
  (5) Derived Distributions
      Standardized values
      χ^2 (Chi-squared) distribution
      t-Distribution
      F-Distribution

CHAPTER 7: SAMPLING THEORY AND METHODS

7.1 THEORY OF SAMPLING
    Random sampling
    Non-random sampling
    Sampling error
    Sampling bias
7.2 RANDOM SAMPLING METHODS
  (1) Simple Random Sampling
      Estimate

11.2 TWO INDEPENDENT SAMPLE NON-PARAMETRIC TESTS
    (1) Median Test
    (2) Mann–Whitney U-test
        Hypotheses
        Mann–Whitney U-test table
        Test statistic

11.3 TWO RELATED SAMPLE NON-PARAMETRIC TESTS
    (1) McNemar's Test
        Hypotheses
        McNemar's test table
        Test statistic
    (2) Sign Test
        Hypotheses
        Sign test table
        Test statistic

11.4 K-INDEPENDENT SAMPLE NON-PARAMETRIC TESTS
    (1) Extended Median Test
    (2) Krushkal–Wallis H-test
        Hypotheses
        Krushkal–Wallis H-test table
        Test statistic

11.5 K-RELATED SAMPLE NON-PARAMETRIC TESTS
    (1) Cochran Q-test
        Hypotheses
        Cochran Q-test table
        Test statistic
    (2) Friedman Test
        Hypotheses
        Friedman test table
        Test statistic

11.6 LOG-LINEAR MODELS
    Introduction
    Two-dimensional tables
    An illustration
    Variables are not independent

CHAPTER 12: CORRELATION ANALYSIS AND REGRESSION ANALYSIS

12.1 TWO-QUANTITATIVE VARIABLES CORRELATION COEFFICIENTS
    (1) Scatter Diagram

CHAPTER 14: SURVIVAL ANALYSIS AND TIME SERIES ANALYSIS

14.1 SURVIVAL ANALYSIS
    (1)   Cohort Hospital Stay Table Technique
    (2)   Current Hospital Stay Table Technique
14.2 TIME SERIES ANALYSIS
    (1)   Time Series Components
          Trend
          Seasonal variation
          Cyclical component
          Random fluctuations
    (2)   Decomposition of Time Series Components
          Trend
          Seasonal variation
          Cyclical component
          Random fluctuations

CHAPTER 15: MULTIVARIATE STATISTICAL METHODS

15.1 PROFILE TECHNIQUES
    (1)   Three-Variables Profile
15.2 MULTIVARIATE CORRELATION COEFFICIENTS
    (1)   Partial Correlation Coefficient
    (2)   Multiple Correlation Coefficient
15.3 MULTIPLE REGRESSION ANALYSIS
    (1)   Two Independent Variables Regression Analysis
          Fitting two independent variables regression equation
15.4 MULTIVARIATE ANALYSIS OF VARIANCE
    (1)   Two Groups and Two Variables
          Layout
          Hypotheses
          Test statistic
15.5 ANALYSIS OF MULTI-DIMENSIONAL CONTINGENCY TABLES
    (1)   Analysis of Three-Dimensional Contingency Table
          $2 \times 2 \times 2$ contingency table
          Hypotheses
          Test statistic

CHAPTER 16: CLUSTER ANALYSIS

16.1 FEATURES OF CLUSTER ANALYSIS
(1) Formal introduction
   Classification, identification, discriminant analysis
(2) Functions of cluster analysis
   Shedding light of previously made hypothesis
   Prediction
   Other purposes
(3) Historical background
(4) Structure type
(5) Importance in psychiatric research
(6) Cluster analysis and factor analysis
(7) Cluster analysis and discriminant analysis
(8) Limitations

16.2 MEASURES OF SIMILARITY
(1) Association coefficients
(2) Correlation coefficient
(3) Distance measures
   Euclidean distance
   Absolute distance
(4) Choice of measures

16.3 CHOICE OF VARIABLES
(1) Confirmatory with theory
(2) Transformation
(3) Weightage to variables
(4) Number of variables

16.4 HIERARCHICAL METHODS
Dendrograms
(1) Single-linkage method (SLINK)
(2) Complete-linkage method (CLINK)
(3) Average-linkage between merged groups(ALINKB)
(4) Average-linkage within the new group(ALINKW)
(5) Centroid method (CENTROID)
(6) Median method (MEDIAN)
(7) The WARD method (WARD)
(8) General problems of hierarchical agglomerative methods

16.5 NUMBER OF CLUSTERS IN HIERARCHICAL METHODS
(1) Complexity of the problem
(2) Empirical investigations
(3) Factors affecting number of clusters

    (4) Clusters structure and number o clusters

    (5) Analytical methods

16.6 PARTITIONING METHODS

    (1) Nature of partitioning methods

    (2) Methods of initiating clusters

          Starting with seed points

          Selecting appropriate starting partitions

    (3) Methods of reallocating entities

    (4) Forgys method

    (5) Jancey variant method

    (6) General problems in partitioning methods

    (7) Determination of number of clusters

          Point-biserial correlation

    (8) Choice of the methods

16.7   VALIDATION TECHNIQUES

    (1) Method of replication

          Rand Index

    (2) Significance test on variables used to create clusters

    (3) Significance test on external variables

    (4) Method of marker samples

    (5) Association with pre-program assessment test scores

16.8   NUMERICAL DEMONSTRATION

16.9   AN EMPIRICAL CLASSIFICATION OF GOVERNMENT MENTAL HOSPITALS IN INDIA

16.10   PATTERN OF DISTRIBUTION OF MENTAL HEALTH MANPOWER : AN INTERNATIONAL SCENE

16.11   AN EMPIRICAL CLASSIFICATION OF CHILD PSYCHIATRY DISORDERS

    (1) Formal introduction

    (2) Criteria for classification

    (3) Classification of disorders and classification of children

    (4) Issues in classification

    (5) Classification models

          Ideographic approach

          Categorical approach

          Dimensional approach

    (6) Application on real data

16.12   CLUSTER ANALYSIS IN EVALUATION OF PROGRAMS

    (1) A formal introduction

    (2) Evaluation approaches

(2) Type, Width, and Decimal
(3) Value and Label
(4) Further Items
(5) Saving the File

21.3 PROCEDURES FOR RUNNING DATA ANALYSIS USING SPSS

(1) Dataset and Data Editor Format
(2) Plan of Analysis
(3) Main Dialogue Box
(4) Output

21.4 DATA HANDLING USING DATA MENU

(1) Split File
     Main dialogue box
(2) Weight Cases
     Main dialogue box

21.5 DATA HANDLING USING TRANSFORM MENU

(1) Recode
     Main dialogue box

CHAPTER 22: RUNNING DATA ANALYSIS USING SPSS

22.1 RUNNING "PARAMETRIC TESTS OF SIGNIFICANCE" (METHODS IN CHAPTER 9)

(1) Running "One-Sample *t*-Test"
     Main dialogue box
(2) Running "Independent Samples *t*-Test"
     Main dialogue box
(3) Running "Paired-Sample *t*-test"
     Main dialogue box

22.2 RUNNING "ANOVA TESTS" (METHODS IN CHAPTER 10)

(1) Running "One-Way ANOVA"
     Main dialogue box
(2) Running "ANCOVA Test"
     Main dialogue box
(3) Running "Repeated Measures ANOVA"
     Main dialogue box

22.3 RUNNING "NON-PARAMETRIC TESTS OF SIGNIFICANCE" (METHODS IN CHAPTER 11)

(1) Running "Chi-Square Test of Significance"
     Main dialogue box
(2) Running "Run Test"

Main dialogue box

(3) Running "Mann–Whitney U-Test"
Main dialogue box

(4) Running "Krushkal–Wallis H-Test"
Main dialogue box

(5) Running "Friedman Test"
Main dialogue box

22.4 RUNNING "CORRELATION ANALYSIS AND REGRESSION ANALYSIS" (METHODS IN CHAPTER 12)

(1) Running "Pearson Correlation Coefficient"
Main dialogue box

(2) Running "Linear Regression Analysis"
Main dialogue box

22.5 RUNNING "RELIABILITY ANALYSIS" (METHODS IN CHAPTER 13)

(1) Running "Cronbach's Alpha"
Main dialogue box

22.6 RUNNING "SURVIVAL ANALYSIS AND TIME SERIES ANALYSIS" (METHODS IN CHAPTER 14)

(1) Running "Life Tables"
Main dialogue box

(2) Running "Sequence Chart"
Main dialogue box

22.7 RUNNING "MULTIVARIATE STATISTICAL METHODS" (METHODS IN CHAPTER 15)

(1) Running "Partial Correlation Coefficient"
Main dialogue box

(2) Running "Multiple Regression Analysis"
Main dialogue box

22.8 RUNNING "CLUSTER ANALYSIS AND DISCRIMINANT ANALYSIS" (METHODS IN CHAPTERS 16 AND 17)

(1) Running "Hierarchical Cluster Analysis"
Main dialogue box

(2) Running "Linear Discriminant Analysis"
Main dialogue box

22.9 RUNNING "FACTOR ANALYSIS" (METHODS IN CHAPTER 18)

(1) Running "Principal Component Analysis"
Main dialogue box

APPENDIX III: AREA, POPULATION SIZE, AND DENSITY OF POPULATION OF COUNTRIES

Continents/ Countries	Area (Lakh Square Kilometers)	Population (Crores)	Density of Population
WORLD	1343.1	733.0	55
ASIA	320.5	437.1	136
RUSSIA	171.0	14.2	8
OCEANIA	85.4	3.8	5
EUROPE	59.7	60.6	102
AMERICA	402.5	98.1	24
AFRICA	304.0	119.2	39
ASIAN COUNTRIES			
Group 1	**(51.4)**	**(171.6)**	**(334)**
India	32.9	126.0	383
Pakistan	8.0	20.2	254
Afghanistan	6.5	3.3	51
Nepal	1.5	3.1	197
Bhutan	0.4	0.1	20
Bangladesh	1.4	16.7	1085
Sri Lanka	0.7	2.2	339
Maldives	0.003	0.04	1319
Group 2	**(30.8)**	**(22.6)**	**(73)**
Iran	16.5	8.3	50
Iraq	4.4	3.8	87
Turkey	7.8	8.0	102
Syria	1.9	1.8	93
Lebanon	0.1	0.6	599
Cyprus	0.1	0.1	130
Group 3	**(42.0)**	**(8.4)**	**(20)**
Tajikistan	1.4	0.8	58
Turkmenistan	4.9	0.5	11
Kazakhstan	27.3	1.8	7
Kyrgyzstan	2.0	0.6	29
Uzbekistan	4.5	2.9	66
Azerbaijan	0.9	1.0	114

Continents/ Countries	Area (Lakh Square Kilometers)	Population (Crores)	Density of Population
Armenia	0.3	0.3	103
Georgia	0.7	0.5	71
Group 4	**(121.0)**	**171.6)**	**(141)**
China	96.0	137.0	143
China (Hong Kong)	0.01	0.7	6642
China (Macau)	0.0002	0.06	37930
Mongolia	15.6	0.3	2
North Korea	1.2	2.5	208
South Korea	1.0	5.1	511
Japan	3.8	12.7	353
Taiwan	0.4	2.4	652
Philippines	3.0	10.8	342
Group 5	**(42.1)**	**(53.9)**	**(128)**
Myanmar	6.8	5.7	84
Thailand	5.1	6.8	133
Vietnam	3.3	9.5	288
Laos	2.4	0.7	30
Cambodia	1.8	1.6	88
Malaysia	3.3	3.1	94
Singapore	0.01	0.6	8293
Brunei	0.1	0.04	76
Indonesia	19.1	25.8	136
East Timor	0.2	0.1	85
Group 6	**(33.2)**	**(9.0)**	**(27)**
Saudi Arabia	22.5	2.8	13
Yemen	5.3	2.7	52
Oman	3.1	0.3	11
United Arab Emirates	0.8	0.6	71
Qatar	0.1	0.2	194
Bahrain	0.008	0.1	1802
Kuwait	0.2	0.3	159
Israel	0.2	0.8	393

Continents/ Countries	Area (Lakh Square Kilometers)	Population (Crores)	Density of Population
Jordon	0.9	0.8	92
Palestine	0.06	0.4	685
Russia	(171.0)	(14.2)	(8)
OCEANIA COUNTRIES			
Group 1	**(80.1)**	**(2.8)**	**(4)**
Australia	77.4	2.3	3
New Zealand	2.7	0.5	17
Group 2	**(5.3)**	**(1.0)**	**(19)**
Fiji	0.2	0.1	50
Kiribati	0.008	0.01	132
Papua New Guinea	4.6	0.7	15
Palau	0.02	0.002	47
Micronesia	0.01	0.01	149
Marshall Islands	0.002	0.01	405
Samoa	0.03	0.02	70
Solomon Islands	0.3	0.1	22
Tonga	0.01	0.01	143
Tuvalu	0.0003	0.001	422
Vanuatu	0.1	0.03	23
Nauru	0.0002	0.001	457
EUROPEAN COUNTRIES			
Group 1	**(19.9)**	**(30.8)**	**(155)**
Ireland	0.7	0.5	70
United Kingdom	2.4	6.4	265
Germany	3.6	8.1	226
Belgium	0.3	1.1	373
Netherland	0.4	1.7	410
Portugal	0.9	1.1	118
Spain	5.1	4.9	96
France	6.2	6.7	104

Continents/ Countries	Area (Lakh Square Kilometers)	Population (Crores)	Density of Population
Andorra	0.005	0.01	183
Albania	0.3	0.3	105
Monaco	0.00002	0.003	15139
Group 2	(4.9)	(8.7)	(178)
Switzerland	0.4	0.8	198
Austria	0.8	0.9	104
Liechtenstein	0.002	0.004	237
Luxembourg	0.03	0.1	225
Slovakia	0.5	0.5	111
Slovenia	0.2	0.2	97
Italy	3.0	6.2	206
San Marino	0.0006	0.003	546
Malta	0.003	0.04	1314
Vatican City (Small city country with 0.4 Square Kilometers and 1000 population)			
Group 3	(10.5)	(9.7)	(92)
Estonia	0.5	0.1	28
Latvia	0.7	0.2	30
Lithuania	0.7	0.3	44
Poland	3.1	3.9	123
Czech Republic	0.8	1.1	135
Hungary	0.9	1.0	106
Kosovo	0.1	0.2	173
Montenegro	0.2	0.1	47
Macedonia	0.3	0.2	82
Serbia	0.8	0.7	92
Croatia	0.6	0.4	76
Bosnia H	0.5	0.4	75
Greece	1.3	1.1	82
Group 4	(11.9)	(8.7)	(73)
Ukraine	6.0	4.4	73
Belarus	2.1	1.0	46
Moldova	0.3	0.4	104
Bulgaria	1.1	0.7	64

Continents/ Countries	Area (Lakh Square Kilometers)	Population (Crores)	Density of Population
Romania	2.4	2.2	91
Group 5	(12.5)	(2.7)	(22)
Norway	3.2	0.5	16
Sweden	4.5	1.0	22
Finland	3.4	0.6	16
Iceland	1.0	0.03	3
Denmark	0.4	0.6	130
AMERICAN COUNTRIES			
Group 1	(177.4)	(41.6)	(24)
Brazil	85.1	20.6	24
Argentina	27.8	4.4	16
Bolivia	11.0	1.1	10
Uruguay	1.8	0.3	19
Paraguay	4.1	0.7	17
Chile	7.6	1.8	23
Peru	12.9	3.1	24
Ecuador	2.8	1.6	57
Colombia	11.4	4.7	41
Venezuela	9.1	3.1	34
Guyana	2.2	0.1	3
Suriname	1.6	0.1	4
Group 2	(217.7)	(48.1)	(22)
United States of America	98.3	32.3	33
Canada	99.8	3.5	4
Mexico	19.6	12.3	63
Group 3	(5.2)	(4.5)	(85)
Guatemala	1.1	1.5	139
Honduras	1.1	0.9	79
Belize	0.2	0.04	15
Nicaragua	1.3	0.6	46
Costa Rica	0.5	0.5	95
El Salvador	0.2	0.6	292

Continents/ Countries	Area (Lakh Square Kilometers)	Population (Crores)	Density of Population
Panama	0.8	0.4	49
Group 4	**(2.2)**	**(3.8)**	**(173)**
Antigua and Barbuda	0.004	0.01	211
Barbados	0.004	0.03	678
Bahamas	0.1	0.03	24
Cuba	1.1	1.1	101
Haiti	0.3	1.1	378
Jamaica	0.1	0.3	270
St. Kitts and Nevis	0.003	0.01	201
St. Lucia	0.006	0.02	267
St. Vincent and Grenadines	0.004	0.01	263
Dominica	0.008	0.01	98
Dominica Republic	0.5	1.1	218
Grenada	0.004	0.01	323
Trinidad and Tobago	0.05	0.1	238
AFRICAN COUNTRIES			
Group 1	**(30.8)**	**(15.5)**	**(50)**
Somalia	6.4	1.1	17
Kenya	5.8	4.7	81
Tanzania	9.5	5.2	55
Malawi	1.2	1.9	157
Mozambique	7.9	2.6	32
Group 2	**(50.9)**	**(12.5)**	**(25)**
South Africa	12.2	5.4	45
Lesotho	0.3	0.2	64
Swaziland	0.2	0.2	84
Botswana	5.8	0.2	4
Namibia	8.2	0.3	3
Angola	12.5	2.0	16

Continents/ Countries	Area (Lakh Square Kilometers)	Population (Crores)	Density of Population
Burundi	0.3	1.1	399
Zambia	7.5	1.6	21
Zimbabwe	3.9	1.5	37
Group 3	**(41.1)**	**(21.0)**	**(51)**
Ethiopia	12.0	10.2	93
Rwanda	0.3	1.3	493
Uganda	2.4	3.8	159
Djibouti	0.2	0.1	37
Eritrea	1.2	0.6	50
Sudan	18.6	3.7	20
South Sudan	6.4	1.3	19
Group 4	**(60.2)**	**(18.7)**	**(31)**
Egypt	10.0	9.5	95
Libya	17.6	0.7	4
Algeria	23.8	4.0	17
Tunisia	1.6	1.1	68
Morocco	4.5	3.4	76
Sahrawi Arab Democratic	2.7	0.04	2
Group 5	**(90.4)**	**(33.1)**	**(37)**
Mauritania	10.3	0.4	4
Mali	12.4	1.8	14
Chad	12.8	1.2	9
Nigeria	9.2	18.6	201
Niger	12.7	1.9	15
Central African Republic	6.2	0.6	9
Congo Republic	3.4	0.5	14
Congo Democratic Republic	23.4	8.1	35

Continents/ Countries	Area (Lakh Square Kilometers)	Population (Crores)	Density of Population
Group 6	(8.4)	(4.6)	(55)
Ghana	2.4	2.7	113
Gambia	0.1	0.2	177
Gabon	2.7	0.2	7
Guinea	2.5	1.2	47
Guinea-Bissau	0.4	0.2	48
Equatorial Guinea	0.3	0.1	27
Group 7	(16.3)	(11.2)	(69)
Cote d'Ivoire	3.2	2.4	74
Cape Verde	0.04	0.06	137
Senegal	2.0	1.4	73
Sierra Leone	0.7	0.6	77
Togo	0.6	0.8	137
Liberia	1.1	0.4	39
Benin	1.1	1.1	95
Burkina Faso	2.7	2.0	71
Cameroon	4.8	2.4	51
São Tomé and Príncipe	0.01	0.02	205
Group 8	(5.9)	(2.6)	(44)
Mauritius	0.02	0.1	657
Madagascar	5.9	2.4	42
Seychelles	0.004	0.1	205
Comoros	0.02	0.08	356

APPENDIX IV: AN EMPIRICAL CLASSIFICATION OF CHILD PSYCHIATRIC DISORDERS

1: General Information

Figures in %

General Information	Total (#435)	Childhood Psychosis (#79)	Hysterical Syndromes (#111)	Anxiety Disorders (#76)	Conduct Disorders (#60)	Hyper-activity (#50)	Scholastic Backward-ness (#59)
Mean age (in years)	(10.6)	11.5**	11.2*	10.8	10.2	8.5	10.2
Males	(65)	60	60	59	73	78*	66
Religion							
Hindu	(85)	75	90	92	82	92	76
Muslim	(12)	23**	7	4	15	8	15
Christian	(3)	2	3	4	3	-	9*
Referral							
General practitioner	(10)	14	12	7	10	8	7
Special doctor	(37)	30	51**	38	30	38	20
Family	(36)	39	20	47*	38	38	42
School	(6)	4	3	3	8	6	14**
Child complaints							
Precipitating factors	(19)	20	38**	18	10	2	5
Acute onset	(30)	18	87**	20	3	4	-
Episodic course	(7)	3	18**	8	-	2	-
Duration of illness							
Upto 3 months	(31)	38	70**	26	7	2	-
3 months – 1 year	(15)	20	20	11	20	12	5
Above 1 year	(54)	42	10	63	73**	86**	95**
With previous treatment	(27)	19	52**	33	12	16	10

Note: * P < 0.05,　　**p < 0.01

2: Symptoms

Figures in %

Symptoms	Total (#435)	Childhood Psychosis (#79)	Hysterical Syndromes (#111)	Anxiety Disorders (#76)	Conduct Disorders (#60)	Hyper-activity (#50)	Scholastic Back-wardness (#59)
Speech / language disorders							
Stammering	(6)	14**	4	7	-	2	7
Lisping	(2)	1	-	1	2	6*	3
Under talkative	(2)	3	-	4	-	-	5*
Immature grammatical structure	(1)	-	-	1	-	-	5**
Reading difficulty							
In own language	(12)	5	-	4	7	30**	44**
In foreign language	(12)	6	-	4	7	28**	42**
Spelling mistakes	(11)	1	-	1	8	28**	42**
Sleep disturbances							
Disturbed sleep	(11)	38**	5	4	7	8	2
Initial insomnia	(3)	14**	-	3	-	2	-
Nightmares / terrors	(3)	5	1	5	2	-	2
Somnambulism	(2)	4	1	1	2	-	3
Excessive reported dreams	(1)	6**	-	-	-	-	-
Appetite							
Abnormal	(12)	37**	4	9	7	10	3
Food fads	(1)	-	-	1	3*	2	-
Problem of elimination							
Enurosis	(10)	13	7	11	13	12	9
Encopresis	(1)	-	-	1	1	4*	-

Note: * P < 0.05, **p < 0.01

2: Symptoms (Continued)

Figures in %

Symptoms	Total (#435)	Childhood Psychosis (#79)	Hysterical Syndromes (#111)	Anxiety Disorders (#76)	Conduct Disorders (#60)	Hyper-activity (#50)	Scholastic Backward-ness (#59)
Behavioral disorders							
Stubborn	(53)	24	54	32	93**	98**	37
Demanding	(38)	15	26	16	90**	98**	15
Disobedient	(31)	8	7	11	97**	98**	12
Temper tantrum	(25)	4	8	7	75**	78**	12
Timid	(20)	37**	9	36**	5	8	25
Distractibility	(18)	15	1	3	20	92**	9
Impulsivity	(17)	8	1	3	32**	88**	3
Inattention	(16)	14	-	1	13	92**	7
Fidgety	(16)	9	-	8	18	80**	7
Overactive	(13)	13	-	-	5	90**	-
Nervous	(12)	30**	3	21*	3	8	9
Aggressive	(11)	4	-	1	42**	38**	2
Bullying others	(11)	3	1	-	42**	34**	2
Loner	(7)	5	-	9	3	18**	12
Antisocial activity	(6)	-	-	3	27**	14*	3
Getting bullied	(5)	5	-	4	2	12*	10
Wandering	(5)	14**	-	4	7	16*	-
Loitering	(4)	3	-	4	12**	12*	2
Motor Retardation	(1)	4*	-	-	-	-	2

Note: * $P < 0.05$,   **$p < 0.01$

2: Symptoms (Continued)

Figures in %

Symptoms	Total (#435)	Childhood Psychosis (#79)	Hysterical Syndromes (#111)	Anxiety Disorders (#76)	Conduct Disorders (#60)	Hyper-activity (#50)	Scholastic Backward-ness (#59)
Mood disorders							
Irritability	(8)	9	2	7	20**	12	5
Worrying	(8)	15*	3	12	8	4	7
Unduly depressed	(6)	18**	4	7	5	-	2
Apprehension / fear	(6)	18**	-	7	5	2	2
Free floating anxiety	(3)	10**	-	1	-	-	3
Unduly happy	(2)	11**	-	-	-	1	-
Hostility	(2)	3	-	-	10*	2	2
Panic	(2)	9**	1	3	-	-	-
Misery	(1)	3	1	3	-	-	-
Loneliness	(1)	1	-	3	-	2	2
Suspicious-ness	(1)	1	-	3	5*	-	-
Suicidal ideation	(2)	4	4	1	2	2	-
Situation-specific anxiety / panic							
Fear of Meeting new people	(4)	8	2	5	-	-	7
Being left alone	(4)	6	3	7	-	2	5
Being in a group	(2)	1	1	5*	-	-	3
Animals	(1)	1	-	4**	-	-	-
Crowds	(1)	4**	-	1	-	-	-

Note: * $P < 0.05$,   **$p < 0.01$

2: Symptoms (Continued)

Figures in %

Symptoms	Total (#435)	Childhood Psychosis (#79)	Hysterical Syndromes (#111)	Anxiety Disorders (#76)	Conduct Disorders (#60)	Hyper-activity (#50)	Scholastic Backward-ness (#59)
School							
Scholastic back-wardness	(44)	14	20	30	62**	84**	98**
Below average intelligence	(22)	1	4	4	10	50**	93**
Irregular to school	(16)	23	13	5	33**	14	9
Abnormal peer group adjust-ment	(13)	5	-	8	18	52**	19
Truancy	(5)	1	-	1	23*	10	-
School phobia	(2)	1	-	3	2	4	2
Neurotic disorders							
Nail biting	(3)	4	1	4	2	10*	3
Tics	(2)	5*	1	1	2	-	2
Thumb sucking	(1)	-	-	1	2	6**	-
Cruelty to animals	(1)	-	-	-	2	10**	-
Hysterical							
Fits	(13)	1	48**	4	2	-	-
Unconscious-ness	(9)	1	26**	7	3	-	-
Motor symp-toms	(6)	1	16**	7	5	-	-
Sensory symp-toms	(6)	8	9	13**	-	-	-
Visceral symp-toms	(2)	1	4	3	-	-	2
Possession	(1)	1	2	-	-	2	-
Amnesia	(1)	1	2	-	-	-	-

Note: * P < 0.05,   **p < 0.01

2: Symptoms (Continued)

Figures in %

Symptoms	Total (#435)	Childhood Psychosis (#79)	Hysterical Syndromes (#111)	Anxiety Disorders (#76)	Conduct Disorders (#60)	Hyper-activity (#50)	Scholastic Backward-ness (#59)
Psychotic symptoms							
Talk to self	(5)	22**	2	-	2	2	-
Laughs to self	(4)	17**	1	1	3	2	-
Increased socialization	(3)	14**	1	-	2	-	-
Cries to self	(2)	8**	1	-	2	2	-
Irrelevant talk	(2)	6**	2	-	-	-	-
Withdrawal	(2)	9**	1	1	-	-	-
Silly smile	(1)	4**	-	-	2	-	-
Neglect personal hygiene	(1)	4**	-	-	2	-	-
Bizare moments	(1)	4**	-	-	-	-	-
Physical complaints							
Headache	(18)	27*	19	29*	3	16	10
Giddiness	(9)	11	14*	13	-	2	3
Epilepsy	(9)	4	14	4	13	22**	2
Pain abdomen	(4)	5	5	1	5	6	2
Menstrual complaints	(1)	4*	1	-	2	-	-

Note: * $P < 0.05$,   **$p < 0.01$

3: Family Atmosphere

Figures in %

Family At-mosphere	Total (#435)	Childhood Psychosis (#79)	Hysterical Symptoms (#111)	Anxiety Disorders (#76)	Conduct Disorders (#60)	Hyper-Activity (#50)	Scholastic Backward-ness (#59)
Family type							
Nuclear	(73)	77	74	74	78	60	73
Extended nuclear	(16)	10	19	17	15	22	12
Joint	(11)	13	7	9	7	18*	15
Mean family size	(6.2)	6.8*	6.4	5.6	5.7	5.9	6.4
Consan-guinity	(26)	25	25	30	32	22	24
Mean family income	(Rs.838)	480	520	1488**	690	890	1034
Family occupation							
Skilled	(35)	53**	33	9	47	30	42
Business	(18)	20	18	18	22	8	15
Service	(16)	5	11	32**	17	22	17
Profes-sional	(14)	3	7	40**	3	18	15
Father							
Mean age	(43.3)	43.7	44.3	44.1	42.4	39.0	44.6
Mean pa-ternal age	(32.7)	32.2	33.1	33.3	32.2	30.4	34.4
Education							
Illiterate	(8)	14**	11	-	5	12	3
Primary, secondary	(66)	82**	79**	26	75	54	71
Graduate, profes-sional	(26)	4	10	74**	20	34	26
Abnormal physical health	(13)	13	16	11	15	10	9
Abnormal mental health	(8)	9	11	4	13	4	3

Note: * $P < 0.05$,　　**$p < 0.01$

3: Family Atmosphere (Continued)

Figures in %

Family Atmosphere	Total (#435)	Childhood Psychosis (#79)	Hysterical Symptoms (#111)	Anxiety Disorders (#76)	Conduct Disorders (#60)	Hyper-Activity (#50)	Scholastic Backward-ness (#59)
Mother							
Mean age	(35.0)	35.3	35.2	35.7	35.2	31.6	35.9
Mean maternal age	(24.4)	23.8	24.0	24.9	25.0	23.0	25.7
Education							
Illiterate	(19)	33*	26	2	17	16	15
Primary, secondary	(71)	67	71	69	70	70	78
Graduate, professional	(10)	-	3	29**	13	14	7
Abnormal physical health	(10)	9	14	9	8	8	7
Abnormal mental health	(8)	15*	6	4	12	8	-
Family atmosphere							
Mental disturbance in other family member	(23)	30	28	20	27	20	10
Discordant intra-familial relationship	(23)	25	20	21	35*	24	14
Lack of warmth intra-familial relationship	(4)	5	3	1	5	6	5
Familial over involvement	(42)	23	44	50	58*	46	34
Inadequate parental control	(24)	9	15	17	70**	28	17
Inadequate living condition	(5)	10*	7	-	2	2	3
Anomalies family situation	(7)	8	10	3	8	6	9
Stress / disturbances in school	(25)	23	41**	25	23	2	19
Migration or social transplantation	(2)	1	1	5*	-	-	3

Note: * $P < 0.05$,   **$p < 0.01$

4: Personal History, School, Family, etc.

Figures in %

Personal History, etc.	Total (#435)	Childhood Psychosis (#79)	Hysterical Symptoms (#111)	Anxiety Disorders (#76)	Conduct Disorders (#60)	Hyper-Activity (#50)	Scholastic Backward-ness (#59)
Personal history							
Abnormal pregnancy	(3)	3	2	1	2	4	5
Abnormal delivery	(8)	4	6	7	10	16*	12
Abnormal birth weight	(3)	1	1	1	7	8*	3
Abnormal birth cry	(3)	1	-	4	7	4	9*
Hyperpyrexia	(1)	-	-	1	-	-	5**
Convulsions	(1)	-	-	3	-	-	5**
Milestones of development							
Delay in motor developments	(10)	4	-	5	5	16*	41**
Delay in speech	(11)	4	1	9	8	24*	49**
School and study pattern							
Mean age of joining	(4.8)	4.9	5.0	4.8	4.9	4.7	4.8
Mean present class	(5.1)	6.2**	5.8**	8.7*	4.5	2.9	3.9
Difficulty at starting school	(18)	8	9	11	20	32*	44**
Change in the medium of instruction	(23)	15	22	26	20	26	34*
Strained relation with the teacher	(37)	15	22	15	67**	88**	53*
Help in studies at home	(26)	5	11	30	22	32	75**

Note: * P < 0.05,　　**p < 0.01

4: Personal History, School, Family, etc. (Continued)

Figures in %

Personal History, etc.	Total (#435)	Childhood Psychosis (#79)	Hysterical Symptoms (#111)	Anxiety Disorders (#76)	Conduct Disorders (#60)	Hyper-Activity (#50)	Scholastic Backward-ness (#59)
Abnormality in premorbid treatment							
Social relation	(8)	6	1	11	2	26**	10
Intellectual activity	(18)	1	2	4	10	46**	75
Mood	(4)	3	1	5	2	18**	3
Attitude to work	(14)	-	3	3	7	54**	41**
Interpersonal relation	(15)	9	6	13	5	46**	27*
Self critical	(12)	1	5	5	8	46**	25**
Complaints	(24)	8	24	8	35*	62**	24

Note: * $P < 0.05$, **$p < 0.01$

5: Mental Status Examination

Figures in %

Mental Status Examination	Total (#435)	Childhood Psychosis (#79)	Hysterical Symptoms (#111)	Anxiety Disorders (#76)	Conduct Disorders (#60)	Hyper-Activity (#50)	Scholastic Backward-ness (#59)
General appearance	(5)	14**	5	3	-	-	2
Level of motor activity	(15)	24*	2	1	5	74**	2
Attention and concentration	(14)	20	1	-	3	80**	-
Poor memory	(1)	-	-	1	-	4*	2
Below average intelligence	(15)	1	1	1	3	36**	75
Mood	(18)	42**	10	20	20	6	7
Thoughts and verbalization	(13)	39**	3	13	3	8	12
Senses and perception	(3)	11**	1	1	-	-	-
Self-image	(8)	24**	-	7	13	2	5
Feeling about family members	(23)	25	26	16	37**	14	14
Perception of his own problem	(46)	44	65**	34	38	72**	14

Note: * $P < 0.05$, **$p < 0.01$

APPENDIX V: PROBABILITY DISTRIBUTION TABLES

Area Under the Standard Normal Curve (Z-values)

Normal Deviate (Z)	.00	.02	.04	.06	.08
0.0	.000	.008	.016	.024	.032
0.1	.040	.048	.056	.064	.071
0.2	.079	.087	.095	.103	.110
0.3	.118	.126	.133	.141	.148
0.4	.155	.163	.170	.177	.184
0.5	.192	.199	.205	.212	.219
0.6	.226	.232	.239	.245	.252
0.7	.258	.264	.270	.276	.282
0.8	.288	.294	.299	.305	.311
0.9	.316	.321	.326	.332	.337
1.0	.341	.346	.351	.355	.360
1.1	.364	.369	.373	.377	.381
1.2	.385	.389	.393	.396	.400
1.3	.403	.407	.410	.413	.416
1.4	.419	.422	.425	.428	.431
1.5	.433	.436	.438	.441	.443
1.6	.445	.447	.450	.452	.454
1.7	.455	.457	.459	.461	.463
1.8	.464	.466	.467	.469	.470
1.9	.471	.473	.474	.475	.476
2.0	.477	.478	.479	.480	.481
2.1	.482	.483	.484	.485	.486
2.2	.486	.487	.488	.488	.489
2.3	.489	.490	.490	.491	.491
2.4	.492	.492	.493	.493	.493
2.5	.494	.494	.495	.495	.495
2.6	.495	.496	.496	.496	.496
2.7	.496	.497	.497	.497	.497
2.8	.497	.498	.498	.498	.498
2.9	.498	.498	.498	.499	.499
3.0	.499	.499	.499	.499	.499

Critical Values of χ^2

Degrees of Freedom (df)	Significance Level (0.05)	Significance Level (0.01)
1	3.84	6.64
2	5.99	9.21
3	7.81	11.35
4	9.49	13.28
5	11.07	15.09
6	12.59	16.81
7	14.07	18.48
8	15.51	20.09
9	16.92	21.67
10	18.31	23.21
11	19.68	24.73
12	21.03	26.22
13	22.36	27.69
14	23.68	29.14
15	25.00	30.58
16	26.30	32.00
17	27.59	33.41
18	28.87	34.81
19	30.14	36.19
20	31.41	37.57
21	32.67	38.93
22	33.92	40.29
23	35.17	41.64
24	36.42	42.98
25	37.65	44.31
26	38.89	45.64
27	40.11	46.96
28	41.34	48.28
29	42.56	49.59
30	43.77	50.89

Critical Values of *t* for Two-tailed Tests

Degrees of Freedom (df)	Significance Level (0.05)	Significance Level (0.01)
1	12.71	63.66
2	4.30	9.93
3	3.18	5.84
4	2.78	4.60
5	2.57	4.03
6	2.45	3.71
7	2.37	3.50
8	2.31	3.36
9	2.26	3.25
10	2.23	3.17
11	2.20	3.11
12	2.18	3.06
13	2.16	3.01
14	2.15	2.98
15	2.13	2.95
16	2.12	2.92
17	2.11	2.90
18	2.10	2.88
19	2.09	2.86
20	2.09	2.85
21	2.08	2.83
22	2.07	2.82
23	2.07	2.81
24	2.06	2.80
25	2.06	2.79
26	2.06	2.78
27	2.05	2.77
28	2.05	2.76
29	2.05	2.75
30	2.04	2.74

Critical Values of F at 5% Level of Significance

df_1 / df_2	1	2	3	4	5	6	7	8
1	161.00	200.00	216.00	225.00	230.00	234.00	237.00	239.00
2	18.51	19.00	19.16	19.25	19.30	19.33	19.35	19.37
3	10.13	9.55	9.28	9.12	9.01	8.94	8.89	8.85
4	7.71	6.94	6.59	6.39	6.26	6.16	6.09	6.04
6	5.99	5.14	4.76	4.53	4.39	4.28	4.21	4.15
8	5.32	4.46	4.07	3.84	3.69	3.58	3.50	3.44
10	4.96	4.10	3.71	3.48	3.33	3.22	3.14	3.07
12	4.75	3.89	3.49	3.26	3.11	3.00	2.91	2.85
14	4.60	3.74	3.34	3.11	2.96	2.85	2.76	2.70
16	4.49	3.63	3.24	3.01	2.85	2.74	2.66	2.59
18	4.41	3.55	3.16	2.93	2.77	2.66	2.58	2.51
20	4.35	3.49	3.10	2.87	2.71	2.60	2.51	2.45
30	4.17	3.32	2.92	2.69	2.53	2.42	2.33	2.27
40	4.08	3.23	2.84	2.61	2.45	2.34	2.25	2.18
50	4.03	3.18	2.79	2.56	2.40	2.29	2.20	2.13
60	4.00	3.15	2.76	2.53	2.37	2.25	2.17	2.10

Critical Values of F at 1% Level of Significance

df_a	1	2	3	4	5	6	7	8
df_b								
1	4048.	4993.	5377.	5577.	5668.	5924.	5992.	6096.
2	98.50	99.01	99.15	99.23	99.30	99.33	99.35	99.39
3	34.12	30.82	29.46	28.71	28.24	27.91	27.67	27.49
4	21.20	18.00	16.69	15.98	15.52	15.21	14.98	14.80
6	13.75	10.92	9.78	9.15	8.75	8.47	8.26	8.10
8	11.26	8.65	7.59	7.01	6.63	6.37	6.18	6.03
10	10.04	7.56	6.55	5.99	5.64	5.39	5.20	5.06
12	9.33	6.93	5.95	5.41	5.06	4.82	4.64	4.50
14	8.86	6.51	5.56	5.04	4.69	4.46	4.28	4.14
16	8.53	6.23	5.29	4.77	4.44	4.20	4.03	3.89
18	8.29	6.01	5.09	4.58	4.25	4.01	3.84	3.71
20	8.10	5.85	4.94	4.43	4.10	3.87	3.70	3.56
30	7.56	5.39	4.51	4.02	3.70	3.47	3.30	3.17
40	7.31	5.18	4.31	3.83	3.51	3.29	3.12	2.99
50	7.17	5.06	4.20	3.72	3.41	3.19	3.02	2.89
60	7.08	4.98	4.13	3.65	3.34	3.12	2.95	2.82
df_a Degree of freedom of the numerator								
df_b Degree of freedom of the denominator								

INDEX